Problems and Solutions for Undergraduate Real Analysis II

by Kit-Wing Yu, PhD

kitwing@hotmail.com

ISBN: 978-988-78797-6-3 (eBook)
ISBN: 978-988-78797-7-0 (Paperback)

About the author

Dr. Kit-Wing Yu received his B.Sc. (1st Hons), M.Phil. and Ph.D. degrees in Math. at the HKUST, PGDE (Mathematics) at the CUHK. After his graduation, he has joined United Christian College to serve as a mathematics teacher for at least nineteen years. He has also taken the responsibility of the mathematics panel since 2002. Furthermore, he was appointed as a part-time tutor (2002 – 2005) and then a part-time course coordinator (2006 – 2010) of the Department of Mathematics at the OUHK.

Apart from teaching, Dr. Yu has been appointed to be a marker of the HKAL Pure Mathematics and HKDSE Mathematics (Core Part) for over thirteen years. Between 2012 and 2014, Dr. Yu was invited to be a Judge Member by the World Olympic Mathematics Competition (China). In the area of academic publication, he is the author of four books

- *A Complete Solution Guide to Real and Complex Analysis I.*

- *Problems and Solutions for Undergraduate Real Analysis I.*

- *Mock Tests for the ACT Mathematics.*

- *A Complete Solution Guide to Principles of Mathematical Analysis.*

Besides, he has published over twelve research papers in international mathematical journals, including some well-known journals such as J. Reine Angew. Math., Proc. Roy. Soc. Edinburgh Sect. A and Kodai Math. J.. His research interests are inequalities, special functions and Nevanlinna's value distribution theory.

Preface

This book *"Problems and Solutions for Undergraduate Real Analysis II"* is the continuum of the first book *"Problems and Solutions for Undergraduate Real Analysis I"*. Its aim is the same as its first book: We want to assist undergraduate students or first-year students who study mathematics in learning their first rigorous real analysis course.

"The only way to learn mathematics is to do mathematics." – Paul Halmos. My learning and teaching experience has convinced me that this assertion is definitely true. In fact, I believe that "doing mathematics" means a lot to everyone who studies or teaches mathematics. It is not only a way of writing a solution to a mathematical problem, but also a mean of reflecting mathematics deeply, exercising mathematical techniques expertly, exchanging mathematical thoughts with others effectively and searching new mathematical ideas unexpectedly. Thus I hope everyone who is reading this book can experience and acquire the above benefits eventually.

The wide variety of problems, which are of varying difficulty, include the following topics: Sequences and Series of Functions, Improper Integrals, Lebesgue Measure, Lebesgue Measurable Functions, Lebesgue Integration, Differential Calculus of Functions of Several Variables and Integral Calculus of Functions of Several Variables. Furthermore, the main features of this book are listed as follows:

- The book contains 226 problems, which cover the topics mentioned above, with *detailed* and *complete* solutions. Particularly, we include over 100 problems for the Lebesgue integration theory which, I believe, is totally new to all undergraduate students.

- Each chapter starts with a brief and concise note of introducing the notations, terminologies, basic mathematical concepts or important/famous/frequently used theorems (without proofs) relevant to the topic.

- Three levels of difficulty have been assigned to problems:

Symbol	Level of difficulty	Meaning
$\star$	Introductory	These problems are basic and every student must be familiar with them.
$\star$ $\star$	Intermediate	The depth and the complexity of the problems increase. Students who target for higher grades must study them.
$\star$ $\star$ $\star$	Advanced	These problems are very difficult and they may need some specific skills.

- Colors are used frequently in order to highlight or explain problems, examples, remarks, main points/formulas involved, or show the steps of manipulation in some complicated proofs. (ebook only)

If you find any typos or mistakes, please feel free to send your valuable comments or opinions to

kitwing@hotmail.com

Any updated errata of this book or news about my new book will be posted on my new website:

https://sites.google.com/view/yukitwing/

Kit Wing Yu
July 2019

List of Figures

Contents

CHAPTER $\boldsymbol{10}$

Sequences and Series of Functions

10.1 Fundamental Concepts

In Chapters 5 and 6, we consider sequences and series of real numbers. In this chapter, we study sequences and series whose terms are **functions**. In fact, the representation of a function as the limit of a sequence or an infinite series of functions arises naturally in advanced analysis. The main references in this chapter are [2, Chap. 9], [4, Chap. 8], [22, Chap. 3] and [28, Chap. 16].

10.1.1 Pointwise and Uniform Convergence

Definition 10.1 (Pointwise Convergence). *Suppose that $\{f_n\}$ is a sequence of real-valued or complex-valued functions on a set $E \subseteq \mathbb{R}$. For every $x \in E$, if the sequence $\{f_n(x)\}$ converges, then the function f defined by the equation*

$$f(x) = \lim_{n \to \infty} f_n(x) \tag{10.1}$$

*is called the **limit function** of $\{f_n\}$ and we can say that $\{f_n\}$ **converges pointwise** to f on E. Similarly, if the series $\sum f_n(x)$ converges pointwise for every $x \in E$, then we can define the function*

$$f(x) = \sum_{n=1}^{\infty} f_n(x) \tag{10.2}$$

*on E and it is called the **sum of the series**.*

Definition 10.2 (Uniform Convergence). *A sequence $\{f_n\}$ of functions defined on $E \subseteq \mathbb{R}$ is said to **converge uniformly** to f on E if for every $\epsilon > 0$, there exists an $N(\epsilon) \in \mathbb{N}$ such that $n \geq N(\epsilon)$ implies that*

$$|f_n(x) - f(x)| < \epsilon \tag{10.3}$$

*for all $x \in E$. In this case, f is called the **uniform limit** of $\{f_n\}$ on E. Similarly, we say that the series $\sum f_n(x)$ **converges uniformly** to f on E if the sequence $\{s_n\}$ of its partial sums, where*

$$s_n(x) = \sum_{k=1}^{n} f_k(x),$$

converges uniformly to f on E.

1

Figure 10.1 shows the sequence of functions $\{f_n\}$ on $[0, 1)$, where $f_n(x) = x^n$. This sequence of functions converges pointwise but *not* uniformly to $f = 0$ because if we take $\epsilon = \frac{1}{2}$, then for each $n \in \mathbb{N}$, one can find a x_n such that $f_n(x_n) > \frac{1}{2}$, i.e., points above the red dotted line. See also Problem 10.4 below.

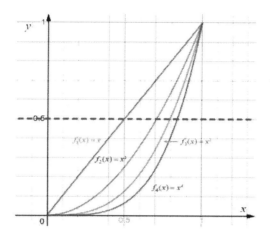

Figure 10.1: An example of pointwise convergence.

Now Figure 10.2 gives the sequence of functions $\{f_n\}$ on $[0, 1]$, where $f_n(x) = \frac{1}{n^2 + x^2}$. In fact, the idea of the inequality (10.3) can be "seen" in the figure that the graphs of all f_n for $n \geq N$ lie "below" the line $y = \epsilon$.

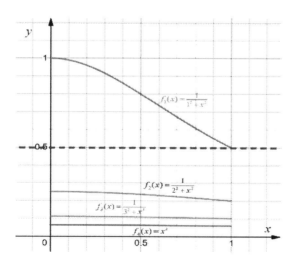

Figure 10.2: An example of uniform convergence.

> **Remark 10.1**
>
> (a) It is obvious that the uniform convergence of $\{f_n\}$ to f implies its pointwise convergence to f, but the converse is false.
>
> (b) Some books use $f_n \to f$ and $f_n \rightrightarrows f$ on E to denote pointwise convergence and uniform convergence respectively.

The core interest for the convergence problem here is that we want to determine what kinds of properties of functions f_n that will be "preserved" under the limiting processes (10.1) or (10.2). In fact, we discover that uniform convergence preserves continuity, differentiation and integrability of the functions f_n.

10.1.2 Criteria for Uniform Convergence

In the following, we state some methods of testing whether a sequence $\{f_n\}$ converges to its pointwise limit f uniformly.

Theorem 10.3 (Cauchy Criterion for Uniform Convergence). *Suppose that $\{f_n\}$ is a sequence of functions defined on $E \subseteq \mathbb{R}$.*

(a) *The sequence $\{f_n\}$ converges uniformly on E if and only if for every $\epsilon > 0$, there is an $N(\epsilon) \in \mathbb{N}$ such that $m, n \geq N(\epsilon)$ imply that*

$$|f_n(x) - f_m(x)| < \epsilon$$

for all $x \in E$.

(b) *The series $\sum f_n$ converges uniformly on E if and only if for every $\epsilon > 0$, there is an $N(\epsilon) \in \mathbb{N}$ such that $m > n \geq N(\epsilon)$ imply that*

$$\left| \sum_{k=n}^{m} f_k(x) \right| < \epsilon$$

for all $x \in E$.

Theorem 10.4. *The sequence $\{f_n\}$ converges uniformly to f on $E \subseteq \mathbb{R}$ if and only if*

$$M_n = \sup_{x \in E} |f_n(x) - f(x)| \to 0$$

as $n \to \infty$.

Theorem 10.5 (Weierstrass M-test). *Suppose that $\{f_n\}$ is a sequence of functions defined on $E \subseteq \mathbb{R}$ and $\{M_n\}$ is a sequence of nonnegative numbers such that $|f_n(x)| \leq M_n$ for all $n = 1, 2, 3, \ldots$. If $\sum M_n$ converges, then $\sum f_n$ converges uniformly on E.*

10.1.3 Preservation Theorems

Theorem 10.6 (Uniform Convergence and Continuity). *Suppose that $f_n \to f$ uniformly on $E \subseteq \mathbb{R}$. If every f_n is continuous at $p \in E$, then the uniform limit f is also continuous at p.*

Theorem 10.7 (Uniform Convergence and Riemann-Stieltjes Integration). *Let $a < b$. Suppose that α is monotonically increasing on $[a,b]$, all $f_n \in \mathscr{R}(\alpha)$ on $[a,b]$ and $f_n \to f$ uniformly on $[a,b]$. Then we have $f \in \mathscr{R}(\alpha)$ on $[a,b]$ and*

$$\int_a^b f \, d\alpha = \int_a^b \lim_{n \to \infty} f_n \, d\alpha = \lim_{n \to \infty} \int_a^b f_n \, d\alpha.$$

In particular, if the series

$$\sum_{n=1}^{\infty} f_n \to f$$

uniformly to f on $[a,b]$, then we have

$$\int_a^b f \, d\alpha = \int_a^b \sum_{n=1}^{\infty} f_n \, d\alpha = \sum_{n=1}^{\infty} \int_a^b f_n \, d\alpha.$$

Theorem 10.8 (Uniform Convergence and Differentiability). *Suppose that $\{f_n\}$ is a sequence of differentiable functions on $[a,b]$, where $a < b$. We suppose that*

(a) *$\{f_n(p)\}$ converges for at least one point $p \in [a,b]$, and*

(b) *$\{f_n'\}$ converges uniformly on $[a,b]$.*

Then there exists a function f defined on $[a,b]$ such that $f_n \to f$ uniformly on $[a,b]$ and

$$f'(x) = \lim_{n \to \infty} f_n'(x) \tag{10.4}$$

on $[a,b]$.

Furthermore, if the sequences $\{f_n(p)\}$ and $\{f_n'\}$ in parts (a) and (b) are replaced by the series $\sum f_n(p)$ and $\sum f_n'$ respectively, then we have the same conclusion and the equation (10.4) is replaced by

$$f'(x) = \sum_{n=1}^{\infty} f_n'(x).$$

10.1.4 Uniformly Boundedness and Equicontinuity

Now we want to find any similarity between sequences of numbers $\{x_n\}$ and sequences of functions $\{f_n\}$. In fact, we have two important and basic questions. The first question origins from the Bolzano-Weierstrass Theorem ([25, Problem 5.25]): Every bounded sequence of real (or complex in fact) sequence has a convergent subsequence. *Is there any similar result for sequences of functions?* To answer this question, we have to define two kinds of boundedness first.

Definition 10.9 (Pointwise Boundedness and Uniformly Boundedness). *Suppose that $\{f_n\}$ is a sequence of functions defined on $E \subseteq \mathbb{R}$.*

(a) *It is said that $\{f_n\}$ is **pointwise bounded** on E if for every $x \in E$, there is a finite-valued function $g : E \to \mathbb{R}$ such that*

$$|f_n(x)| < g(x)$$

for all $x \in E$ and $n = 1, 2, \ldots$.

(b) It is said that $\{f_n\}$ is **uniformly bounded** on E if there is a positive constant M such that

$$|f_n(x)| < M$$

for all $x \in E$ and $n = 1, 2, \ldots$.

> ### Remark 10.2
>
> It is clear from the definitions that uniformly boundedness certainly implies pointwise boundedness.

It is well-known that if $\{f_n\}$ is **pointwise bounded** on a *countable* set E, then it has a subsequence $\{f_{n_k}\}$ such that $\{f_{n_k}(x)\}$ converges for every $x \in E$.[a] This answers the first question partially. However, the general situation *does not* hold even if E is compact and $\{f_n\}$ is a uniformly bounded sequence of continuous functions.

Next, the second question origins from the fact that a sequence $\{x_n\}$ converges if and only if every subsequence $\{x_{n_k}$ converges (see [25, Theorem 5.3, p. 50]). *Is every convergent sequence of functions contains a uniformly convergent subsequence?* The answer is negative even if we assume stronger conditions that $\{f_n\}$ is uniformly bounded on a compact set E.

Both failures of the above two questions are due to the lack of the concept of **equicontinuity** which is stated as follows:

Definition 10.10 (Equicontinuity). *Let X be a metric space with metric d and $\mathscr{F}$ a family of functions defined on $E \subseteq X$. The family $\mathscr{F}$ is **equicontinuous** on E if for every $\epsilon > 0$, there exists a $\delta > 0$ such that*

$$|f(x) - f(y)| < \epsilon$$

for all $f \in \mathscr{F}$ and all $x, y \in X$ with $d(x, y) < \delta$.

Theorem 10.11. *Suppose that K is a compact metric space. If $\{f_n\}$ is a sequence of continuous functions on K and it converges uniformly on K, then $\{f_n\}$ is equicontinuous on K.*

The Arzelà-Ascoli Theorem. *Suppose that K is a compact metric space and $\{f_n\}$ is a sequence of continuous functions on K. If $\{f_n\}$ is pointwise bounded and equicontinuous on K, then it is uniformly bounded on K. Furthermore, $\{f_n\}$ contains a uniformly convergent subsequence $\{f_{n_k}\}$ on K.*

▎ 10.1.5 The Space $\mathscr{C}(X)$ and the Approximation by Polynomials

Definition 10.12. *Suppose that X is a metric space. Then $\mathscr{C}(X)$ denotes the set of all complex-valued bounded continuous functions on X. For each $f \in \mathscr{C}(X)$, we define*

$$\|f\| = \sup_{x \in X} |f(x)|$$

*which is called the **supremum norm**. Then $\mathscr{C}(X)$ is a complete metric space with this metric $\|\cdot\|$.*

---

[a] See, for example, [18, Theorem 7.23, p. 156].

> **Remark 10.3**
>
> If $\mathscr{A}$ is closed in $\mathscr{C}(X)$, then $\mathscr{A}$ is call **uniformly closed** and its closure $\overline{\mathscr{A}}$ is called **uniform closure**.

By Theorem 10.6 (Uniform Convergence and Continuity) or Problem 10.15, we know that the uniform limit of a sequence of polynomials is a continuous function. It is natural to ask its converse: Can a continuous function be approximated *uniformly* by polynomials? The answer to this question is affirmative and in fact, we state it as follows:

The Weierstrass Approximation Theorem. *If $f : [a,b] \to \mathbb{C}$ is continuous, then there exists a sequence of polynomials $\{P_n\}$ such that $P_n \to f$ uniformly on $[a,b]$. If f is real, then we may take every P_n to be real.*

10.2 Uniform Convergence for Sequences of Functions

> **Problem 10.1**
>
> $(\star)$ *For each $n \in \mathbb{N}$, we define*
>
> $$f_n(x) = \begin{cases} \frac{x}{n}, & \text{if } n \text{ is odd;} \\ \\ \frac{1}{n}, & \text{otherwise.} \end{cases}$$
>
> *Prove that $\{f_n\}$ converges pointwise but not uniformly on $\mathbb{R}$.*

Proof. For every $x \in \mathbb{R}$, since

$$\lim_{k \to \infty} f_{2k+1}(x) = \lim_{k \to \infty} \frac{x}{2k+1} = 0 \quad \text{and} \quad \lim_{k \to \infty} f_{2k}(x) = \lim_{k \to \infty} \frac{1}{2k} = 0,$$

we conclude that $\{f_n\}$ converges pointwise to $f = 0$ on $\mathbb{R}$. However, since

$$M_{2k+1} = \sup_{x \in \mathbb{R}} |f_{2k+1}(x) - 0| = \sup_{x \in \mathbb{R}} \left| \frac{x}{2k} \right| = +\infty,$$

Theorem 10.4 says that $\{f_n\}$ does not converge uniformly on $\mathbb{R}$. This completes the proof of the problem. ∎

> **Problem 10.2**
>
> $(\star)$ *Verify that the sequence of functions $f_n(x) = \frac{x^2+nx}{n}$ converges uniformly on $[0,1]$, but not on $\mathbb{R}$.*

Proof. Obviously, we have $f_n(x) \to x$ as $n \to \infty$ for every $x \in \mathbb{R}$, so its pointwise limit is $f(x) = x$. Now for each *fixed* $n \in \mathbb{N}$, we have

$$M_n = \sup_{x \in \mathbb{R}} |f_n(x) - f(x)| = \sup_{x \in \mathbb{R}} \frac{x^2}{n} = \infty,$$

so Theorem 10.4 implies that $\{f_n\}$ does not converge uniformly to $f(x) = x$ on $\mathbb{R}$. However, on the interval $[0, 1]$, we note that

$$M_n = \frac{1}{n}$$

for every $n \in \mathbb{N}$ so that $M_n \to 0$ as $n \to \infty$. Hence it deduces from Theorem 10.4 again that $\{f_n\}$ converges uniformly to $f(x) = x$ on $[0, 1]$. We have completed the proof of the problem. ∎

Problem 10.3 (Dini's Theorem)

⭐ *Suppose that $\{f_n\}$ is a sequence of continuous functions defined on $[a, b]$ such that $f_n(x) \geq f_{n+1}(x)$ for all $x \in [a, b]$ and $n \in \mathbb{N}$. If $\{f_n\}$ converges pointwise to a continuous function f on $[a, b]$, prove that $f_n \to f$ uniformly on $[a, b]$.*

Proof. Without loss of generality, we assume that $f = 0$. Since each f_n is uniformly continuous on $[a, b]$, the number

$$M_n = \sup_{x \in [a,b]} |f_n(x)|$$

is well-defined. Since $f_n(x) \geq f_{n+1}(x)$ on $[a, b]$, $\{M_n\}$ is decreasing. If M_n does not converge to 0, then there exists a $\epsilon > 0$ such that $M_n > \epsilon$ for all $n \in \mathbb{N}$. This means that for each n, we can find a $x_n \in [a, b]$ with

$$f_n(x_n) > \epsilon. \tag{10.5}$$

Since $\{x_n\} \subseteq [a, b]$, it follows from the Bolzano-Weierstrass Theorem [25, Problem 5.25, p. 68] that $\{x_n\}$ has a convergent subsequence. Let the limit of the subsequence be p. By the hypothesis, we have $f_n(p) \to 0$ as $n \to \infty$. In other words, there is an $N \in \mathbb{N}$ such that

$$f_N(p) < \epsilon.$$

Since f_N is continuous at p, the Sign-preserving Property [25, Problem 7.15, p. 112] ensures that there is a $\delta > 0$ such that

$$f_N(x) < \epsilon \tag{10.6}$$

for all $x \in [a, b]$ with $|x - p| < \delta$. Now the property $f_n(x) \geq f_{n+1}(x)$ on $[a, b]$ implies that the inequality (10.6) also holds for all f_n with $n \geq N$. By the definition of p, we can choose x_n with $n \geq N$ and $|x_n - p| < \delta$ so that the inequality (10.6) gives

$$f_n(x_n) < \epsilon \tag{10.7}$$

for all $n \geq N$. Now it is obvious that the inequalities (10.5) and (10.7) are contrary. Hence we see that $M_n \to 0$ as $n \to \infty$, completing the proof of the problem. ∎

Problem 10.4

⭐ *Show that the hypothesis "f is continuous" in Problem 10.3 (Dini's Theorem) cannot be dropped.*

Proof. Let $f_n(x) = x^n$ on $[0,1]$. Then it is easy to see that $f_n(1) = 1$ for all $n \in \mathbb{N}$ and $f_n(x) \to 0$ as $n \to \infty$ for all $x \in [0,1)$. Thus the limit function f of this sequence $\{f_n\}$ is given by

$$f(x) = \begin{cases} 1, & \text{if } x = 1; \\ \\ 0, & \text{otherwise.} \end{cases} \tag{10.8}$$

Assume the convergence was uniform. Then Theorem 10.6 (Uniform Convergence and Continuity) shows that the function f must also be continuous on $[0,1]$, but it contradicts the definition (10.8). Thus the condition "f is continuous" in Problem 10.3 (Dini's Theorem) cannot be dropped, completing the proof of the problem. ∎

Problem 10.5

$(\star)$ *Construct a sequence of continuous functions $\{f_n\}$ defined on a compact set K with pointwise continuous limit function f but it does not converge uniformly to f on K.*

Proof. Suppose that $K = [0,1]$ and each $f_n : [0,1] \to \mathbb{R}$ is defined by

$$f_n(x) = \begin{cases} nx, & \text{if } x \in [0, \frac{1}{n}]; \\ \\ 2 - nx, & \text{if } x \in (\frac{1}{n}, \frac{2}{n}]; \\ \\ 0, & \text{otherwise.} \end{cases}$$

Then it is clear that every f_n is continuous on $[0,1]$. If $x = 0$, then $f_n(0) = 0$ for every $n \in \mathbb{N}$. If $x \in (0,1]$, then there exists an $N \in \mathbb{N}$ such that $x > \frac{2}{N}$ so that $x \notin (0, \frac{2}{N}]$ and thus $f_n(x) = 0$ for every $n \geq N$. In other words, its limit function is

$$f = 0$$

which is continuous on $[0,1]$. However, we note that for each $n \in \mathbb{N}$, we have

$$M_n = \sup_{x \in [0,1]} |f_n(x) - 0| = f_n\left(\frac{1}{n}\right) = 1$$

and we follow from Theorem 10.4 that $\{f_n\}$ does not converge to $f = 0$ uniformly. This is the end of the proof. ∎

Problem 10.6

$(\star)$ *Suppose $f_n \to f$ uniformly on $E \subseteq \mathbb{R}$ and there is a positive constant M such that $|f_n(x)| \leq M$ on E and all $n \in \mathbb{N}$. Let g be continuous on $[-M, M]$. Define $h = g \circ f$ and $h_n = g \circ f_n$ for every $n \in \mathbb{N}$. Show that $h_n \to h$ uniformly on E.*

Proof. Since g is continuous on $[-M, M]$, we know from [25, Theorem 7.10, p. 100] that g is uniformly continuous on $[-M, M]$. Given $\epsilon > 0$, there is a $\delta > 0$ such that

$$|g(x) - g(y)| < \epsilon \tag{10.9}$$

for all $x, y \in [-M, M]$ with $|x - y| < \delta$. Since $f_n \to f$ uniformly on E, there is an $N \in \mathbb{N}$ such that $n \geq N$ implies that

$$|f_n(x) - f(x)| < \delta \tag{10.10}$$

for every $x \in E$. Combining the inequalities (10.9) and (10.10), it can be shown that $n \geq N$ implies

$$|h_n(x) - h(x)| = |g(f_n(x)) - g(f(x))| < \epsilon$$

for every $x \in E$. By Definition 10.2 (Uniform Convergence), $\{h_n\}$ converges uniformly to h on E. We complete the proof of the problem. ∎

Problem 10.7

$(\star)$ Define $f_n : [0, 1] \to \mathbb{R}$ by

$$f_n(x) = \frac{1}{1 + x^n}.$$

Prove that $\{f_n\}$ converges uniformly to f on $[0, \alpha]$, but not on $[0, 1]$, where $0 < \alpha < 1$.

Proof. Let f be the pointwise limit function of $\{f_n\}$. If $x = 1$, then $f_n(1) = \frac{1}{2}$ for every $n \in \mathbb{N}$. Thus we have $f(1) = \frac{1}{2}$. Let $x \in [0, 1)$. Since

$$\lim_{n \to \infty} f_n(x) = \lim_{n \to \infty} \frac{1}{1 + x^n} = 1,$$

we obtain $f(x) = 1$ on $x \in [0, 1)$. In other words, we have

$$f(x) = \begin{cases} \frac{1}{2}, & \text{if } x = 1; \\ 1, & \text{otherwise.} \end{cases} \tag{10.11}$$

Next, we fix $\alpha \in (0, 1)$. For every $x \in [0, \alpha]$, we have $x^n \leq \alpha^n$ so that $x^n + x^n \alpha^n \leq \alpha^n + x^n \alpha^n$. By this and the definition (10.11), we know that

$$|f_n(x) - f(x)| = \left| \frac{1}{1 + x^n} - 1 \right| = \frac{x^n}{1 + x^n} \leq \frac{\alpha^n}{1 + \alpha^n}. \tag{10.12}$$

Since $0 < \alpha < 1$, we have

$$\lim_{n \to \infty} \frac{\alpha^n}{1 + \alpha^n} = 0.$$

Therefore, we conclude from this and the inequality (10.12) that $f_n \to f$ uniformly on $[0, \alpha]$.

However, given a positive integer n, we let $x \in (\sqrt[n]{\frac{1}{2}}, 1)$ so that $\frac{1}{2} < x^n < 1$. By this and the definition (10.11) again, it yields that

$$|f_n(x) - f(x)| = \left| \frac{1}{1 + x^n} - 1 \right| = \frac{x^n}{1 + x^n} > \frac{\frac{1}{2}}{1 + 1} = \frac{1}{4}$$

and then $\{f_n\}$ does not converge to f on $[0, 1]$ by Theorem 10.4. Hence we complete the proof of the problem. ∎

Problem 10.8

$(\star)$ For $n \in \mathbb{N}$, we suppose that $f_n = \frac{1}{n}\exp(-n^2x^2)$ on $\mathbb{R}$. Prove that $f_n \to 0$ uniformly on $\mathbb{R}$, $f_n' \to 0$ pointwise on $\mathbb{R}$, but not uniformly on $(-M, M)$ for every $M > 0$.

Proof. For every $x \in \mathbb{R}$, note that $e^{x^2} \geq 1$ and so

$$|f_n(x)| = \frac{1}{ne^{n^2x^2}} \leq \frac{1}{n}.$$

By Theorem 10.4, $f_n \to 0$ uniformly on $\mathbb{R}$. By direct differentiation, we have $f_n'(x) = -2nxe^{-n^2x^2}$, so it is easy to see that $f_n' \to 0$ pointwise on $\mathbb{R}$.

Assume that there was a $M > 0$ such that $f_n' \to 0$ uniformly on $(-M, M)$. Pick $\epsilon = e^{-1}$. By assumption, there exists an $N \in \mathbb{N}$ such that $n \geq N$ implies that

$$|f_n'(x)| = \frac{2n|x|}{e^{n^2x^2}} < e^{-1} \tag{10.13}$$

on $(-M, M)$. Now we may take N to be large enough so that $N > \frac{1}{M}$, i.e., $\frac{1}{N} \in (0, M)$. Then it is legal to put $x = \frac{1}{N}$ into the inequality (10.13) to get the contradiction that

$$2e^{-1} < e^{-1}.$$

Hence $\{f_n'\}$ does not converge uniformly to $f' = 0$ on $(-M, M)$, completing the proof of the problem. ∎

Problem 10.9

$(\star)$ Suppose that $\{f_n\}$ is a sequence of uniformly continuous functions on $\mathbb{R}$. If $f_n \to f$ uniformly on $\mathbb{R}$, prove that f is uniformly continuous on $\mathbb{R}$.

Proof. Given $\epsilon > 0$. By the hypotheses, there exists an $N \in \mathbb{N}$ such that $n \geq N$ implies that

$$|f_n(x) - f(x)| < \frac{\epsilon}{3} \tag{10.14}$$

on $\mathbb{R}$. Since f_N is uniformly continuous on $\mathbb{R}$, there exists a $\delta > 0$ such that

$$|f_N(x) - f_N(y)| < \frac{\epsilon}{3} \tag{10.15}$$

for every pair $x, y \in \mathbb{R}$ with $|x - y| < \delta$. Combining the inequalities (10.14) and (10.15), we see immediately that

$$|f(x) - f(y)| \leq |f(x) - f_N(x)| + |f_N(x) - f_N(y)| + |f_N(y) - f(y)| < \epsilon$$

for all $x, y \in \mathbb{R}$ with $|x - y| < \delta$. Hence f is uniformly continuous on $\mathbb{R}$ which completes the proof of the problem. ∎

Problem 10.10

$(\star)$ Let $\{f_n\}$ be a sequence of bounded functions on $E \subseteq \mathbb{R}$. If $f_n \to f$ uniformly on E, prove that f is bounded on E. Is it true for pointwise convergence on bounded E?

Proof. Given $\epsilon > 0$. Then one can find an $N \in \mathbb{N}$ such that $n \geq N$ implies that

$$|f_n(x) - f(x)| < \epsilon.$$

For this N, since f_N is bounded on E, there is a $M > 0$ such that $|f_N(x)| \leq M$ on E. Thus we obtain

$$|f(x)| \leq |f(x) - f_N(x)| + |f_N(x)| < \epsilon + M$$

on E, i.e., f is bounded on E.

The second assertion is negative. For example, we consider $E = (0, 1)$ and

$$f_n(x) = \min(x^{-2}, n).$$

Then it is trivial to check that $f_n(x) \to x^{-2}$ pointwise on $(0, 1)$. Since x^{-2} is unbounded on the bounded set $(0, 1)$, this counterexample completes the proof of the problem. ∎

Problem 10.11

$(\star)$ Prove that

$$\lim_{n \to \infty} \int_1^\pi e^{-nx^2}\, \mathrm{d}x = 0.$$

Proof. Consider $f_n(x) = e^{-nx^2}$ defined on $[1, \pi]$. Clearly, we have $f_n \in \mathscr{R}$ on $[1, \pi]$ because f_n is continuous on $[1, \pi]$ and

$$M_n = \sup_{x \in [1, \pi]} |f_n(x)| \leq \frac{1}{e^n} \to 0$$

as $n \to \infty$. Thus $\{f_n\}$ converges uniformly to 0 on $[1, \pi]$ by Theorem 10.4. By Theorem 10.7 (Uniform Convergence and Riemann-Stieltjes Integration), we see easily that

$$\lim_{n \to \infty} \int_1^\pi e^{-nx^2}\, \mathrm{d}x = \int_1^\pi \lim_{n \to \infty} e^{-nx^2}\, \mathrm{d}x = 0$$

which completes the proof of the problem. ∎

Problem 10.12

$(\star)$ Let $f_n : \mathbb{R} \to \mathbb{R}$. Prove that if $\{f_n\}$ converges pointwise to f on $\mathbb{R}$, then it converges uniformly to f on any finite subset of $\mathbb{R}$.

Proof. Let $\{x_1, x_2, \ldots, x_m\}$ be a finite subset of $\mathbb{R}$. Given $\epsilon > 0$. Since $f_n \to f$ pointwise on $\mathbb{R}$, there exists an $N(\epsilon, x_k) \in \mathbb{N}$ such that $n \geq N(\epsilon, x_k)$ implies that

$$|f_n(x_k) - f(x_k)| < \epsilon, \tag{10.16}$$

where $k = 1, 2, \ldots, m$. Take

$$N(\epsilon) = \max(N(\epsilon, x_1), N(\epsilon, x_2), \ldots, N(\epsilon, x_m)).$$

Then it is obvious that the inequality (10.16) holds for all $k = 1, 2, \ldots, m$ when $n \geq N(\epsilon)$. In other words, $\{f_n\}$ converges uniformly to f on $\{x_1, x_2, \ldots, x_m\}$ which ends the proof of the problem. ∎

Problem 10.13

$(\star)(\star)$ Let $f_n : \mathbb{R} \to \mathbb{R}$. Prove that if $\{f_n\}$ converges uniformly to f on all countable subsets of $\mathbb{R}$, then $\{f_n\}$ converges uniformly to f on $\mathbb{R}$. Can the hypothesis "all countable subsets" be replaced by the condition "all Cauchy sequences"?

Proof. Assume that $\{f_n\}$ did not converge uniformly to f on $\mathbb{R}$. In other words, *for some $\epsilon > 0$* and *for every $N \in \mathbb{N}$, there exists* a $x \in \mathbb{R}$ such that

$$|f_n(x) - f(x)| \geq \epsilon$$

for some $n \geq N$. Particularly, pick $N = 1$, then one can find a $x_1 \in \mathbb{R}$ such that

$$|f_n(x_1) - f(x_1)| \geq \epsilon$$

hold for some $n \geq 1$. In fact, we can find a countable subset $E = \{x_k\}$ of $\mathbb{R}$ such that

$$|f_n(x_k) - f(x_k)| \geq \epsilon \tag{10.17}$$

hold for some $n \geq k$, where $k = 1, 2, \ldots$. However, the inequality (10.17) means that $\{f_n\}$ *does not* converge uniformly on the countable set E, a contradiction.

The second assertion is false. For example, we consider

$$f_n(x) = \begin{cases} 0, & \text{if } x \leq n - \frac{1}{2}; \\ x - n + \frac{1}{2}, & \text{if } n - \frac{1}{2} \leq x \leq n; \\ \frac{1}{2}, & \text{otherwise.} \end{cases} \tag{10.18}$$

Let $\{p_k\}$ be a Cauchy sequence in $\mathbb{R}$. We claim that $f_n \to 0$ uniformly on $\{p_k\}$. We know from [25, Theorem 5.2, p. 49] that $\{p_k\}$ is bounded by a positive constant M. On the one hand, if $n > M + \frac{1}{2}$, then $p_k < n - \frac{1}{2}$ for every $k = 1, 2, \ldots$ and we obtain from the definition (10.18) that

$$f_n(p_k) = 0$$

for every $k = 1, 2, \ldots$. Consequently, $f_n \to 0$ uniformly on $\{p_k\}$. On the other hand, for every $n \in \mathbb{N}$, since

$$M_n = \sup_{x \in \mathbb{R}} |f_n(x) - 0| = \frac{1}{2},$$

Theorem 10.4 implies that $\{f_n\}$ does not converge uniformly on $\mathbb{R}$. This completes the proof of the problem.

∎

Problem 10.14

⋆ ⋆ *Suppose that $\{f_n\}$ is a sequence of continuous functions defined on a compact set $K \subset \mathbb{R}$, $f_n \to f$ pointwise on K and f is continuous. Prove that $f_n \to f$ uniformly on K if and only if for every $\epsilon > 0$, there is a $m \in \mathbb{N}$ and a $\delta > 0$ such that $n > m$ and $|f_k(x) - f(x)| < \delta$ imply that*

$$|f_{k+n}(x) - f(x)| < \epsilon \tag{10.19}$$

for all $x \in K$ and $k \in \mathbb{N}$.

Proof. Suppose that $f_n \to f$ uniformly on K. Thus given $\epsilon > 0$, there exists an $N \in \mathbb{N}$ such that $n > N$ implies that

$$|f_n(x) - f(x)| < \epsilon$$

for all $x \in K$. If we take $m = N$ and $\delta = \epsilon$, then the inequality (10.19) holds trivially.

Conversely, given $\epsilon > 0$. By the hypotheses, there exists a $m \in \mathbb{N}$ and a $\delta > 0$ such that $n > m$ and $|f_k(x) - f(x)| < \delta$ imply the inequality (10.19) on K and all $k \in \mathbb{N}$. Fix $p \in K$. Since $f_n(p) \to f(p)$ as $n \to \infty$, there exists a $k \in \mathbb{N}$ such that

$$|f_k(p) - f(p)| < \frac{\delta}{3}. \tag{10.20}$$

Recall that both f_k and f are continuous at p, so there exists a neighborhood $B(p, r_p)$ of radius $r_p > 0$ such that $x \in B(p, r_p) \cap K$ implies

$$|f_k(x) - f_k(p)| < \frac{\delta}{3} \quad \text{and} \quad |f(x) - f(p)| < \frac{\delta}{3}. \tag{10.21}$$

Combining the inequalities (10.20) and (10.21), we see that

$$|f_k(x) - f(x)| \le |f_k(x) - f_k(p)| + |f_k(p) - f(p)| + |f(p) - f(x)| < \delta \tag{10.22}$$

for all $x \in B(p, r_p) \cap K$. It is clear that $\{B(p, r_p) \,|\, p \in K\}$ is an open cover of K, so there exists a finite index $\{p_1, p_2, \ldots, p_s\}$ such that

$$K \subseteq B(p_1, r_{p_1}) \cup B(p_2, r_{p_2}) \cup \cdots \cup B(p_s, r_{p_s}).$$

Suppose that k_i is the corresponding positive integer satisfying the inequality (10.22) for the point p_i, where $i = 1, 2, \ldots, s$. Then for arbitrary $x \in K$, we have $x \in B(p_i, r_{p_i})$ *for some i* so that $x \in B(p_i, r_{p_i}) \cap K$ and thus

$$|f_{k_i}(x) - f(x)| < \delta.$$

By our hypotheses, if $n > m$, then we have

$$|f_{k_i+n}(x) - f(x)| < \epsilon \tag{10.23}$$

for all $x \in B(p_i, r_{p_i}) \cap K$. Put $N = \max(k_1, k_2, \ldots, k_s) + m$. If $n' > N \ge k_i + m$ for every $i = 1, 2, \ldots, s$, then we deduce from the inequality (10.23) that

$$|f_{n'}(x) - f(x)| < \epsilon. \tag{10.24}$$

Since x is arbitrary in K, we conclude from the inequality (10.24) that $f_n \to f$ uniformly on K and we complete the proof of the problem. ∎

Problem 10.15

$(\star)\,(\star)$ *Suppose that $\{P_n\}$ is a sequence of polynomials on $\mathbb{R}$ with complex coefficients and $P_n \to P$ uniformly on $\mathbb{R}$. Is P again a polynomial on $\mathbb{R}$?*

Proof. We claim that there exists an $N \in \mathbb{N}$ such that

$$\deg P_n = \deg P_N$$

for all $n \geq N$. Assume that it was not the case. Since $P_n \to P$ uniformly on $\mathbb{R}$, Theorem 10.3 (Cauchy Criterion for Uniform Convergence) ensures that there exists a $k \in \mathbb{N}$ such that $m, n \geq k$ imply that

$$|P_m(x) - P_n(x)| < 1 \tag{10.25}$$

for all $x \in \mathbb{R}$. For this particular k, our assumption shows that there is an $n_k > k$ with the property

$$\deg P_{n_k} \neq \deg P_k$$

and this definitely shows that

$$\sup_{x \in \mathbb{R}} |P_{n_k}(x) - P_k(x)| = \infty.$$

which contradicts the inequality (10.25). Hence this proves the claim.

Now we let $\deg P_N = K$ and

$$P_n(x) = a_{n,K} x^K + a_{n,K-1} x^{K-1} + \cdots + a_{n,0}, \tag{10.26}$$

for all $n \geq N$, where $a_{n,K}, a_{n,K-1}, \ldots, a_{n,0} \in \mathbb{C}$. If $a_{n,i} \neq a_{N,i}$ for some $i \in \{1, 2, \ldots, K\}$, then we have

$$\sup_{x \in \mathbb{R}} |P_n(x) - P_N(x)| = \infty$$

which contradicts the inequality (10.25) again. In other words, we have $a_{n,i} = a_{N,i}$ for all $n \geq N$ and $i = 1, 2, \ldots, K$. For simplicity, we write $a_{N,i} = a_i$ and the polynomial (10.26) can be expressed as

$$P_n(x) = a_K x^K + a_{K-1} x^{K-1} + \cdots + a_1 x + a_{n,0},$$

for all $n \geq N$. If $a_{n,0} \to a_0$ as $n \to \infty$, then the function $P(x)$ must be in the form

$$P(x) = a_K x^K + a_{K-1} x^{K-1} + \cdots + a_1 x + a_0$$

which is a polynomial on $\mathbb{R}$ with complex coefficients. This completes the proof of the problem. ∎

Problem 10.16

$(\star)\,(\star)$ *Suppose that $f_n : \mathbb{R} \to \mathbb{R}$ and all f_n satisfy **Lipschitz condition** with the same **Lipschitz constant** K. If $f_n \to f$ uniformly on $\mathbb{R}$, prove that f also satisfies Lipschitz condition with Lipschitz constant K.*

Proof. For the definitions of Lipschitz condition and Lipschitz constant, you are suggested to read [25, Remark 8.7, p. 142].

Given $\epsilon > 0$, there exists an $N \in \mathbb{N}$ such that $n \geq N$ implies that

$$|f_n(x) - f(x)| < \frac{\epsilon}{2} \tag{10.27}$$

for all $x \in \mathbb{R}$. Now for any pair $x, p \in \mathbb{R}$, we deduce from the inequality (10.27) that

$$|f(x) - f(p)| \leq |f(x) - f_N(x)| + |f_N(x) - f_N(p)| + |f_N(p) - f(p)| < \epsilon + |f_N(x) - f_N(p)|. \tag{10.28}$$

Since f_N satisfies the Lipschitz condition with Lipschitz constant K, it means that

$$|f_N(x) - f_N(p)| \leq K|x - p|$$

for all $x, p \in \mathbb{R}$. Thus the inequality (10.28) gives

$$|f(x) - f(p)| < \epsilon + K|x - p| \tag{10.29}$$

for all $x, p \in \mathbb{R}$. Since ϵ is arbitrary, the inequality (10.29) actually implies that

$$|f(x) - f(p)| \leq K|x - p|$$

for all $x, p \in \mathbb{R}$. Hence f also satisfies Lipschitz condition with Lipschitz constant K, completing the proof of the problem. ∎

10.3 Uniform Convergence for Series of Functions

> **Problem 10.17**
>
> (★) *Let $p > 1$. Show that the series*
>
> $$\sum_{n=1}^{\infty} \left[\frac{\pi}{2} - \arctan[n^p(1 + x^2)] \right]$$
>
> *converges uniformly on $\mathbb{R}$.*

Proof. Recall the fact that $\arctan x + \arctan \frac{1}{x} = \frac{\pi}{2}$ for $x > 0$, so

$$0 < \frac{\pi}{2} - \arctan[n^p(1 + x^2)] = \arctan \frac{1}{n^p(1 + x^2)} \tag{10.30}$$

for all $x \in \mathbb{R}$. Let $f(x) = x - \arctan x$ for $x > 0$. Since $f'(x) = \frac{x^2}{1+x^2} > 0$, f is strictly increasing on $(0, \infty)$ by [25, Remark 8.3, p. 129] and thus

$$x > \arctan x$$

if $x > 0$. Therefore, we further reduce the inequality (10.30) to

$$0 < \frac{\pi}{2} - \arctan[n^p(1 + x^2)] < \frac{1}{n^p(1 + x^2)} \le \frac{1}{n^p}$$

for all $x \in \mathbb{R}$. By [25, Theorem 6.10, p. 77], the series $\sum_{n=1}^{\infty} \frac{1}{n^p}$ converges so that we follow from Theorem 10.5 (Weierstrass M-test) that the series

$$\sum_{n=1}^{\infty} \left[\frac{\pi}{2} - \arctan[n^p(1 + x^2)] \right]$$

converges uniformly on $\mathbb{R}$. This completes the proof of the problem.   ■

Problem 10.18

(⋆) Let $p > \frac{1}{2}$. Prove that the series

$$\sum_{n=1}^{\infty} \frac{x}{n^p(1 + nx^2)}$$

converges uniformly on $\mathbb{R}$.

Proof. For each $n \in \mathbb{N}$, let

$$f_n(x) = \frac{x}{n^p(1 + nx^2)}$$

be defined on $\mathbb{R}$. Using the A.M. $\ge$ G.M., we know that

$$\frac{n^p(1 + nx^2)}{2} = \frac{n^p + n^{p+1}x^2}{2} \ge \sqrt{n^{2p+1}x^2} = n^{p+\frac{1}{2}}|x|$$

which implies

$$\left| \frac{x}{n^p(1 + nx^2)} \right| \le \frac{1}{2n^{p+\frac{1}{2}}}$$

for all $x \in \mathbb{R}$. Since $p > \frac{1}{2}$, it yields from [25, Theorem 6.10, p. 77] that $\sum_{n=1}^{\infty} \frac{1}{n^{p+\frac{1}{2}}}$ converges and Theorem 10.5 (Weierstrass M-test) shows that the series

$$\sum_{n=1}^{\infty} \frac{x}{n^p(1 + nx^2)}$$

converges uniformly on $\mathbb{R}$. We have completed the proof of the problem.   ■

Problem 10.19

$(\star)$ *Suppose that each* $f_n : [0,1] \to (0,\infty)$ *is continuous and*

$$f(x) = \sum_{n=1}^{\infty} f_n(x)$$

is also continuous on $[0,1]$. *Prove that the series* $\sum_{n=1}^{\infty} f_n(x)$ *converges uniformly on* $[0,1]$.

Proof. Let $S_n(x) = \sum_{k=1}^{n} f_k(x)$ and

$$R_n(x) = \sum_{k=n+1}^{\infty} f_k(x) = f(x) - S_n(x),$$

where $n = 1, 2, \ldots$ and $x \in [0,1]$. Since f and S_n are continuous on $[0,1]$, each R_n is continuous on $[0,1]$. By the definition, we also have

$$R_n(x) \geq R_{n+1}(x)$$

for all $x \in [0,1]$ and all $n \in \mathbb{N}$. Furthermore, $R_n(x) \to 0$ pointwise on $[0,1]$. Hence it follows from Problem 10.3 (Dini's Theorem) that $R_n(x) \to 0$ uniformly on $[0,1]$, i.e.,

$$\sum_{k=1}^{n} f_k(x) \to f(x)$$

uniformly on $[0,1]$. This ends the proof of the problem. ∎

Problem 10.20

$(\star)$ *Prove that*

$$f(x) = \sum_{n=1}^{\infty} \frac{x^3 \sin nx}{n^2}$$

is continuous on $\mathbb{R}$.

Proof. Let $N \in \mathbb{N}$. On $(-N, N)$, we have

$$\left| \frac{x^3 \sin nx}{n^2} \right| \leq \frac{N^3}{n^2}.$$

Since $\sum_{n=1}^{\infty} \frac{N^3}{n^2} < \infty$, Theorem 10.5 (Weierstrass M-test) implies that the series

$$f(x) = \sum_{n=1}^{\infty} \frac{x^3 \sin nx}{n^2}$$

converges uniformly on $(-N, N)$.

Suppose that

$$f_n(x) = \sum_{k=1}^{n} \frac{x^3 \sin kx}{k^2}$$

on $(-N, N)$. Then the previous paragraph verifies immediately that $f_n \to f$ uniformly on $(-N, N)$. Since each f_n is continuous on $(-N, N)$, Theorem 10.6 (Uniform Convergence and Continuity) implies that f is continuous on $(-N, N)$. Since N is arbitrary, f is then continuous on $\mathbb{R}$. We have completed the proof of the problem. ∎

Problem 10.21

(⋆)(⋆) *Suppose that $\{a_n\}$ is a sequence of nonzero real numbers. Prove that*

$$\sum_{n=1}^{\infty} \frac{e^{ia_n x}}{n^2}$$

converges uniformly on $\mathbb{R}$ to a continuous function $f : \mathbb{R} \to \mathbb{C}$. Evaluate the limit

$$\lim_{T \to \infty} \frac{1}{2T} \int_{-T}^{T} f(x)\, dx.$$

Proof. Clearly, we have

$$\left| \frac{e^{ia_n x}}{n^2} \right| \leq \frac{1}{n^2}$$

for all $x \in \mathbb{R}$ so that Theorem 10.5 (Weierstrass M-test) implies that the series converges uniformly on $\mathbb{R}$ to a function $f : \mathbb{R} \to \mathbb{C}$. Since the partial sum of the series is obviously continuous on $\mathbb{R}$, Theorem 10.6 (Uniform Convergence and Continuity) ensures that the function f must be continuous on $\mathbb{R}$. Furthermore, since the convergence is uniform, Theorem 10.7 (Uniform Convergence and Riemann-Stieltjes Integration) gives

$$\frac{1}{2T} \int_{-T}^{T} f(x)\, dx = \frac{1}{2T} \int_{-T}^{T} \sum_{n=1}^{\infty} \frac{e^{ia_n x}}{n^2}\, dx = \frac{1}{2T} \sum_{n=1}^{\infty} \int_{-T}^{T} \frac{e^{ia_n x}}{n^2}\, dx = \sum_{n=1}^{\infty} \frac{\sin a_n T}{n^2 \cdot a_n T}. \tag{10.31}$$

By the Mean Value Theorem for Derivatives [25, p. 129], it is true that $|\sin x| \leq |x|$ for all $x \in \mathbb{R}$ so that

$$\left| \frac{\sin a_n T}{n^2 \cdot a_n T} \right| \leq \frac{1}{n^2}$$

for all $n \in \mathbb{N}$ and all $T \in \mathbb{R}$. Next, we follow from Theorem 10.5 (Weierstrass M-test) that the infinite series

$$\sum_{n=1}^{\infty} \frac{\sin a_n T}{n^2 \cdot a_n T}$$

converges uniformly on $\mathbb{R}$ (with respect to T). Therefore, it deduces from the equation (10.31) and then using [18, Exercise 13, p. 198] to conclude that

$$\lim_{T \to \infty} \frac{1}{2T} \int_{-T}^{T} f(x)\, dx = \lim_{T \to \infty} \sum_{n=1}^{\infty} \frac{\sin a_n T}{n^2 \cdot a_n T}$$

$$= \sum_{n=1}^{\infty} \frac{1}{n^2} \Big(\lim_{T \to \infty} \frac{\sin a_n T}{a_n T} \Big)$$

$$= \sum_{n=1}^{\infty} \frac{1}{n^2}$$

$$= \frac{\pi^2}{6},$$

completing the proof of the problem.   ■

Problem 10.22

(⋆) Suppose that $f : \mathbb{R} \to \mathbb{R}$ is continuous and $-\infty < a < b < \infty$. For each $n \in \mathbb{N}$, we define $f_n : \mathbb{R} \to \mathbb{R}$ by

$$f_n(x) = \frac{1}{n} \sum_{k=0}^{n-1} f\Big(x + \frac{k}{n} \Big).$$

Prove that f_n converges uniformly on $[a, b]$.

Proof. Given $\epsilon > 0$. Since f is continuous on $\mathbb{R}$, it is uniformly continuous on $[a, b+1]$. Then there exists a $\delta > 0$ such that

$$|f(x) - f(y)| < \epsilon \tag{10.32}$$

for every $x, y \in [a, b+1]$ and $|x - y| < \delta$. Particularly, we may choose an $n \in \mathbb{N}$ such that $\frac{1}{n} < \delta$. Fix $x \in [a, b]$. Then we have

$$\Big| \int_x^{x+1} f(t)\,dt - f_n(x) \Big| = \Big| \sum_{k=0}^{n-1} \int_{x+\frac{k}{n}}^{x+\frac{k+1}{n}} f(t)\,dt - \frac{1}{n} \sum_{k=0}^{n-1} f\Big(x + \frac{k}{n}\Big) \Big|. \tag{10.33}$$

By the First Mean Value Theorem for Integrals [25, p. 162], for each $k = 1, 2, \ldots, n-1$, there is a $y_k \in (x + \frac{k}{n}, x + \frac{k+1}{n}) \subseteq [a, b+1]$ such that

$$\int_{x+\frac{k}{n}}^{x+\frac{k+1}{n}} f(t)\,dt = \frac{f(y_k)}{n}. \tag{10.34}$$

Substituting the value (10.34) into the equation (10.33), we obtain

$$\Big| \int_x^{x+1} f(t)\,dt - f_n(x) \Big| = \Big| \sum_{k=0}^{n-1} \frac{f(y_k)}{n} - \frac{1}{n} \sum_{k=0}^{n-1} f\Big(x + \frac{k}{n}\Big) \Big|$$

$$\le \frac{1}{n} \sum_{k=0}^{n-1} \Big| f(y_k) - f\Big(x + \frac{k}{n}\Big) \Big|. \tag{10.35}$$

Finally, by using the uniform continuity (10.32) to the inequality (10.35), we see that

$$\Big| \int_x^{x+1} f(t)\,dt - f_n(x) \Big| < \epsilon$$

for all $x \in [a, b]$ and all positive integers $n > \frac{1}{\delta}$. Now if we define $F : [a, b] \to \mathbb{R}$ by

$$F(x) = \int_x^{x+1} f(t) \, \mathrm{d}t,$$

then we conclude immediately from Definition 10.2 (Uniform Convergence) that $\{f_n\}$ converges uniformly to F on $[a, b]$. This completes the proof of the problem. ∎

Problem 10.23

$(\star)(\star)$ *Show that the infinite series*

$$\sum_{n=1}^{\infty} \frac{\sin nx}{n}$$

converges uniformly on $[\delta, 2\pi - \delta]$, where $\delta \in (0, \pi)$.

Proof. Given $\epsilon > 0$. Fix $\delta \in (0, \pi)$. We claim that the series

$$\sum_{k=1}^{\infty} \frac{\mathrm{e}^{ikx}}{k} \tag{10.36}$$

converges uniformly on $[\delta, 2\pi - \delta]$ first. To this end, we note that if $x \in [\delta, 2\pi - \delta]$, then

$$|1 - \mathrm{e}^{ix}| = \sqrt{2(1 - \cos x)} \geq \sqrt{2(1 - \cos \delta)} > 0.$$

Therefore, we have

$$|\mathrm{e}^{inx} + \mathrm{e}^{i(n+1)x} + \cdots + \mathrm{e}^{imx}| = \left| \frac{\mathrm{e}^{inx} - \mathrm{e}^{i(m+1)x}}{1 - \mathrm{e}^{ix}} \right| \leq \sqrt{\frac{2}{1 - \cos \delta}}, \tag{10.37}$$

where $m \geq n \geq 0$. Recall the summation by parts [18, Theorem 3.41, p. 70] that

$$\sum_{k=n}^{m} a_k b_k = A_m b_m - A_{n-1} b_n + \sum_{k=n}^{m-1} A_k (b_k - b_{k+1}), \tag{10.38}$$

where $A_k = a_0 + a_1 + a_2 + \cdots + a_k$. Hence we deduce from the inequality (10.37) and the formula (10.38) with $a_k = \mathrm{e}^{ikx}$ that

$$\left| \sum_{k=n}^{m} \frac{\mathrm{e}^{ikx}}{k} \right| \leq \left| \frac{A_m}{m} - \frac{A_{n-1}}{n} + \sum_{k=n}^{m-1} A_k \left(\frac{1}{k} - \frac{1}{k+1} \right) \right|$$

$$\leq \sqrt{\frac{2}{1 - \cos \delta}} \cdot \left(\frac{1}{m} + \frac{1}{n} \right) + \sqrt{\frac{2}{1 - \cos \delta}} \cdot \sum_{k=n}^{m-1} \left| \frac{1}{k} - \frac{1}{k+1} \right|$$

$$\leq \frac{2}{n} \cdot \sqrt{\frac{2}{1 - \cos \delta}} + \sqrt{\frac{2}{1 - \cos \delta}} \cdot \left(\frac{1}{n} - \frac{1}{m} \right)$$

$$\leq \frac{4}{n} \cdot \sqrt{\frac{2}{1 - \cos \delta}}. \tag{10.39}$$

Consequently, if n is large enough, then we get immediately from the inequality (10.39) that

$$\Big| \sum_{k=n}^{m} \frac{e^{ikx}}{k} \Big| < \epsilon.$$

Thus we conclude from Theorem 10.3 (Cauchy Criterion for Uniform Convergence) that the series (10.36) converges uniformly on $[\delta, 2\pi - \delta]$ as desired. Finally, we get from the definition of the modulus easily that

$$\Big| \sum_{k=n}^{m} \frac{e^{ikx}}{k} \Big| \geq \Big| \sum_{k=n}^{m} \frac{\sin kx}{k} \Big|,$$

so we may obtain the same conclusion and thus we complete the analysis of the problem. ■

Problem 10.24

$\star$ $\star$ *Suppose that $f : \mathbb{R} \to \mathbb{R}$ is given by*

$$f(x) = \sum_{n=1}^{\infty} \frac{1}{n^2 + x^2}.$$

Prove that f is differentiable in $\mathbb{R}$ and find $f'(x)$.

Proof. Let $f_n(x) = \frac{1}{n^2 + x^2}$ and M be a positive constant. Recall that

$$\sum_{n=1}^{\infty} f_n(0) = \sum_{n=1}^{\infty} \frac{1}{n^2} = \frac{\pi^2}{6}.$$

In addition, if $|x| \leq M$, then we have

$$|f_n'(x)| \leq \Big| \frac{-2x}{(n^2 + x^2)^2} \Big| \leq \frac{2M}{n^4}$$

so that

$$\Big| \sum_{n=1}^{\infty} f_n'(x) \Big| \leq \sum_{n=1}^{\infty} |f_n'(x)| \leq 2M \cdot \sum_{n=1}^{\infty} \frac{1}{n^4} < \infty.$$

By Theorem 10.5 (Weierstrass M-test), the series

$$\sum_{n=1}^{\infty} f_n'(x)$$

converges uniformly on $[-M, M]$. Now it deduces from Theorem 10.8 (Uniform Convergence and Differentiability) that f is differentiable in $[-M, M]$ and

$$f'(x) = \sum_{n=1}^{\infty} f_n'(x) = \sum_{n=1}^{\infty} \frac{-2x}{(n^2 + x^2)^2} \qquad (10.40)$$

Since M is arbitrary, the formula (10.40) holds for all $x \in \mathbb{R}$ and we complete the proof of the problem. ■

Problem 10.25

$\star$ $\star$ Let $f(x) = \displaystyle\int_0^x \frac{\sin t}{t}\, dt$, where $x \in \mathbb{R}$. Prove that

$$f(x) = \sum_{n=0}^{\infty} \frac{(-1)^n x^{2n+1}}{(2n+1)!(2n+1)}$$

for $x \in \mathbb{R}$.

Proof. Since the power series representation of $\sin t$ is given by

$$\sin t = \sum_{n=0}^{\infty} \frac{(-1)^n}{(2n+1)!} t^{2n+1},$$

we have

$$\frac{\sin t}{t} = \sum_{n=0}^{\infty} \frac{(-1)^n}{(2n+1)!} t^{2n}. \tag{10.41}$$

Let $M > 0$. Since

$$\left| \frac{(-1)^n t^{2n}}{(2n+1)!} \right| \le \frac{M^{2n}}{(2n+1)!}$$

for $t \in [-M, M]$ and the series

$$\sum_{n=0}^{\infty} \frac{M^{2n}}{(2n+1)!}$$

converges, it yields from Theorem 10.5 (Weierstrass M-test) that the series on the right-hand side of the representation (10.41) converges uniformly on $[-M, M]$. It is definitely that each $\frac{(-1)^n}{(2n+1)!} t^{2n} \in \mathscr{R}$ on $[-M, M]$, it follows from Theorem 10.7 (Uniform Convergence and Riemann-Stieltjes Integration) that

$$\begin{aligned}
f(x) &= \int_0^x \frac{\sin t}{t}\, dt \\
&= \int_0^x \sum_{n=0}^{\infty} \frac{(-1)^n}{(2n+1)!} t^{2n}\, dt \\
&= \sum_{n=0}^{\infty} \int_0^x \frac{(-1)^n}{(2n+1)!} t^{2n}\, dt \\
&= \sum_{n=0}^{\infty} \frac{(-1)^n}{(2n+1)!(2n+1)} x^{2n+1}
\end{aligned}$$

for all $x \in [-M, M]$. Since M is arbitrary, our desired result follows. This completes the proof of the problem. ∎

10.4 Equicontinuous Families of Functions

Problem 10.26

$(\star)(\star)$ *Let* $-\infty < a < b < \infty$. *Suppose that* $\{f_n\}$ *is pointwise convergent and equicontinuous on* $[a, b]$. *Prove that* $\{f_n\}$ *converges uniformly on* $[a, b]$.

Proof. Given $\epsilon > 0$. Let f be the limit function of $\{f_n\}$ on $[a, b]$. By the hypotheses, there is a $\delta > 0$ such that

$$|f_n(x) - f_n(y)| < \frac{\epsilon}{3} \tag{10.42}$$

for all $x, y \in [a, b]$ with $|x - y| < \delta$ and all $n \in \mathbb{N}$. Taking $n \to \infty$ in the inequality (10.42), we get

$$|f(x) - f(y)| \le \frac{\epsilon}{3} \tag{10.43}$$

for all $x, y \in [a, b]$ with $|x - y| < \delta$. By the definition, the inequality (10.43) shows that f is uniformly continuous on $[a, b]$. Since $[a, b]$ is compact, there exists a set of finite points $\{x_1, x_2, \ldots, x_k\} \subseteq [a, b]$ such that

$$[a, b] \subseteq \bigcup_{i=1}^{k} (x_i - \delta, x_i + \delta). \tag{10.44}$$

Since $f_n \to f$ pointwise on $[a, b]$, there exists an $N_i \in \mathbb{N}$ such that $n \ge N_i$ implies

$$|f_n(x_i) - f(x_i)| < \frac{\epsilon}{3}, \tag{10.45}$$

where $i = 1, 2, \ldots, k$. Put $N = \max(N_1, N_2, \ldots, N_k)$. If $x \in [a, b]$, then the finite open covering property (10.44) ensures that there exists an $i \in \{1, 2, \ldots, k\}$ such that $|x - x_i| < \delta$. Therefore, for all $x \in [a, b]$, it follows from the inequalities (10.42), (10.43) and (10.45) that $n \ge N$ implies

$$|f_n(x) - f(x)| \le |f_n(x) - f_n(x_i)| + |f_n(x_i) - f(x_i)| + |f(x_i) - f(x)| < \epsilon.$$

By the definition, $\{f_n\}$ converges uniformly on $[a, b]$. This completes the proof of the problem. ∎

Problem 10.27

$(\star)$ *Let* $-\infty < a < b < \infty$. *Suppose that* $\{f_n\}$ *is a sequence of continuous functions on* $[a, b]$ *and each* f_n *is differentiable in* (a, b). *If* $\{f_n\}$ *is pointwise convergent on* $[a, b]$ *and* $\{f_n'\}$ *is uniformly bounded on* (a, b), *prove that* $\{f_n\}$ *is uniformly convergent on* $[a, b]$.

Proof. For all $x, y \in [a, b]$ and all $n \in \mathbb{N}$, we get from the Mean Value Theorem for Derivatives that

$$|f_n(x) - f_n(y)| = |f_n'(\xi)| \cdot |x - y| \tag{10.46}$$

for some $\xi \in (x, y)$. Since there is a positive constant M such that $|f_n'(x)| \le M$ for all $x \in [a, b]$ and $n \in \mathbb{N}$, the equation (10.46) gives

$$|f_n(x) - f_n(y)| \le M|x - y|.$$

By Definition 10.10 (Equicontinuity), we conclude that $\{f_n\}$ is equicontinuous and our desired result follows immediately from Problem 10.26. This completes the proof of the problem.   ■

Problem 10.28

$(\star)$ Let $-\infty < a < b < \infty$. Suppose that $\{f_n\}$ is a sequence of twice differentiable functions on $[a, b]$. Furthermore, we have $f_n(a) = f'_n(a) = 0$ and $|f''_n(x)| \leq 1$ for all $x \in [a, b]$ and $n \in \mathbb{N}$. Prove that $\{f_n\}$ has a uniformly convergent subsequence on $[a, b]$.

Proof. Applying the Mean Value Theorem for Derivatives to the sequence $\{f'_n\}$, we see that there exists a $\xi \in (x, y)$ such that

$$|f'_n(x) - f'_n(y)| = |f''_n(\xi)| \cdot |x - y| \leq |x - y| \tag{10.47}$$

for all $x, y \in [a, b]$ and all $n \in \mathbb{N}$. Put $y = a$ in the inequality (10.47), we get

$$|f'_n(x)| \leq |x - a| \leq b - a$$

for all $x \in [a, b]$ and all $n \in \mathbb{N}$. Similarly, we apply the Mean Value Theorem for Derivatives to the sequence $\{f_n\}$, so there is a $\theta \in (x, y)$ such that

$$|f_n(x) - f_n(y)| = |f'_n(\theta)| \cdot |x - y| \leq (b - a) \cdot |x - y| \tag{10.48}$$

for all $x, y \in [a, b]$ and all $n \in \mathbb{N}$. Given $\epsilon > 0$, if we take $\delta = \frac{\epsilon}{2(b-a)}$, then it follows from the estimate (10.48) that

$$|f_n(x) - f_n(y)| \leq \frac{\epsilon}{2} < \epsilon$$

holds for all $x, y \in [a, b]$ satisfying $|x-y| < \delta$ and all $n \in \mathbb{N}$. By Definition 10.10 (Equicontinuity), the family $\{f_n\}$ is equicontinuous. By the estimate (10.48) again, if $y = a$, then

$$|f_n(x)| = |f_n(x) - f_n(a)| \leq (b - a) \cdot |x - a| \leq (b - a)^2$$

for all $x \in [a, b]$ and all $n \in \mathbb{N}$. Thus $\{f_n\}$ is uniformly bounded. Hence we follow from the Arzelà-Ascoli Theorem that $\{f_n\}$ has a uniformly convergent subsequence on $[a, b]$, completing the proof of the problem.   ■

Problem 10.29

$(\star)$ Suppose that $\{f_n\}$ is a family of continuously differentiable functions on $[0, 1]$ and it satisfies

$$|f'_n(x)| \leq x^{-\frac{1}{p}}$$

for all $x \in (0, 1]$, where $p > 1$. Furthermore, we have

$$\int_0^1 f_n(x) \, dx = 0. \tag{10.49}$$

Prove that $\{f_n\}$ has a uniformly convergent subsequence.

Proof. For $x, y \in [0, 1]$ with $0 \leq x < y \leq 1$ and $n \in \mathbb{N}$, we have

$$|f_n(x) - f_n(y)| = \left| \int_x^y f_n'(t) \, dt \right| \leq \int_x^y |f_n'(t)| \, dt \leq \int_x^y t^{-\frac{1}{p}} \, dt = \frac{p}{p-1} (y^{1-\frac{1}{p}} - x^{1-\frac{1}{p}}). \quad (10.50)$$

Define the function $g : [0, 1] \to \mathbb{R}$ by $g(x) = x^{1-\frac{1}{p}}$ which is clearly uniformly continuous on $[0, 1]$. In other words, given $\epsilon > 0$, there is a $\delta > 0$ such that

$$|x^{1-\frac{1}{p}} - y^{1-\frac{1}{p}}| = |g(x) - g(y)| < \frac{(p-1)\epsilon}{p} \quad (10.51)$$

for all $x, y \in [0, 1]$ with $|x - y| < \delta$. Combining the inequalities (10.50) and (10.51), we establish that

$$|f_n(x) - f_n(y)| < \epsilon$$

for all $x, y \in [0, 1]$ with $|x - y| < \delta$ and all $n \in \mathbb{N}$. By Definition 10.10 (Equicontinuity), the sequence $\{f_n\}$ is equicontinuous.

By the integral (10.49), it is impossible that $f_n(x) > 0$ or $f_n(x) < 0$ for all $x \in [0, 1]$. In other words, for each positive integer n, this observation and the continuity of f_n guarantee that there exists a $x_n \in [0, 1]$ such that $f_n(x_n) = 0$. Using the estimate (10.50) with $y = x_n$, we have

$$|f_n(x)| \leq \frac{p}{p-1} |x^{1-\frac{1}{p}} - x_n^{1-\frac{1}{p}}| \leq \frac{2p}{p-1}$$

for all $x \in [0, 1]$, i.e., $\{f_n\}$ is uniformly bounded. Now the application of the Arzelà-Ascoli Theorem implies that it contains a uniformly convergent subsequence on $[0, 1]$. We have completed the proof of the problem. ∎

Problem 10.30

⊛ *Suppose that $\{f_n\}$ is a collection of continuous functions on $[0, 1]$. In addition, suppose that $\{f_n\}$ pointwise converges to 0 on $[0, 1]$ and there exists a positive constant M such that*

$$\left| \int_0^1 f_n(x) \, dx \right| \leq M \quad (10.52)$$

for all $n \in \mathbb{N}$. Is it true that

$$\lim_{n \to \infty} \int_0^1 f_n(x) \, dx = 0? \quad (10.53)$$

Proof. The answer is negative. In fact, for every $n \in \mathbb{N}$, we consider $f_n : [0, 1] \to \mathbb{R}$ by

$$f_n(x) = \begin{cases} 2n^2 x, & \text{if } x \in [0, \frac{1}{2n}); \\ -2n^2 x + 2n, & \text{if } x \in [\frac{1}{2n}, \frac{1}{n}); \\ 0, & \text{if } x \in [\frac{1}{n}, 1]. \end{cases}$$

It is easy to check that each f_n is continuous on $[0, 1]$. Now we have

$$f_n(0) = f_n(1) = 0$$

for every $n \in \mathbb{N}$. If $x \in (0,1)$, then one can find an $N \in \mathbb{N}$ such that $x \geq \frac{1}{N}$ which implies that

$$f_n(x) = 0$$

for all $n \geq N$. Therefore, it means that $\{f_n\}$ pointwise converges to 0 on $[0,1]$. Since the graph of the f_n is an isosceles triangle with vertices $(0,0)$, $(\frac{1}{2n}, n)$ and $(\frac{1}{n}, 0)$, we have

$$\int_0^1 f_n(x)\,\mathrm{d}x = \frac{1}{2}$$

which implies the truth of the bound (10.52), but the failure of the limit (10.53). Hence we have completed the proof of the problem.   ∎

Problem 10.31 (Abel's Test for Uniform Convergence)

(⋆)(⋆) *Suppose that $\{g_n\}$ is a sequence of real-valued functions defined on $E \subseteq \mathbb{R}$ such that $g_{n+1}(x) \leq g_n(x)$ for every $x \in E$ and every $n \in \mathbb{N}$. If the family $\{g_n\}$ is uniformly bounded on E and if $\sum f_n$ converges uniformly on E, prove that the series*

$$\sum_{n=1}^{\infty} f_n(x) g_n(x)$$

also converges uniformly on E.

Proof. Since $\{g_n\}$ is uniformly bounded on E, there is a positive constant M such that

$$|g_n(x)| \leq M \qquad\qquad (10.54)$$

for all $x \in E$ and all $n \in \mathbb{N}$. Since $\sum f_n$ converges uniformly on E, given $\epsilon > 0$, there is an $N \in \mathbb{N}$ such that $n > m \geq N$ implies

$$\left| \sum_{k=m}^{n} f_k(x) \right| < \frac{\epsilon}{3M} \qquad\qquad (10.55)$$

for all $x \in E$. Let $F_{m,n}(x) = \sum_{k=m}^{n} f_k(x)$. Then we have

$$\sum_{k=m}^{n} f_k g_k = f_m g_m + f_{m+1} g_{m+1} + \cdots + f_n g_n$$

$$= F_{m,m} g_m + (F_{m,m+1} - F_{m,m}) g_{m+1} + \cdots + (F_{m,n} - F_{m,n-1}) g_n$$
$$= F_{m,m}(g_m - g_{m+1}) + F_{m,m+1}(g_{m+1} - g_{m+2}) + \cdots + F_{m,n-1}(g_{n-1} - g_n)$$
$$+ F_{m,n} g_n. \qquad\qquad (10.56)$$

Since $g_{n+1}(x) \leq g_n(x)$ for every $x \in E$ and every $n \in \mathbb{N}$, we have

$$g_k(x) - g_{k+1}(x) \geq 0$$

for all $x \in E$ and $k = m, m+1, \ldots, n-1$. Combining the estimate (10.55) and the formula (10.56), if $n > m \geq N$, then

$$\left| \sum_{k=m}^{n} f_k(x) g_k(x) \right| \leq |F_{m,m}(x)| \cdot [g_m(x) - g_{m+1}(x)] + |F_{m,m+1}(x)| \cdot [g_{m+1}(x) - g_{m+2}(x)]$$

$$+ \cdots + |F_{m,n-1}(x)| \cdot [g_{n-1}(x) - g_n(x)] + |F_{m,n}(x)| \cdot |g_n(x)|$$

$$< \frac{\epsilon}{3M} \cdot [g_m(x) - g_{m+1}(x)] + \cdots + \frac{\epsilon}{3M} \cdot [g_{n-1}(x) - g_n(x)] + \frac{\epsilon}{3M} \cdot |g_n(x)|$$

$$= \frac{\epsilon}{3M} \cdot [g_m(x) - g_n(x)] + \frac{\epsilon}{3M} \cdot |g_n(x)|. \tag{10.57}$$

Obviously, we have $0 \leq g_m(x) - g_n(x) \leq 2M$. Therefore, we follow from the estimates (10.54) and (10.57) that

$$\left| \sum_{k=m}^{n} f_k(x) g_k(x) \right| < \epsilon$$

for all $x \in E$. By Theorem 10.3 (Cauchy Criterion for Uniform Convergence), we conclude that $\sum f_n(x) g_n(x)$ converges uniformly on E. This completes the analysis of the proof. ∎

Problem 10.32

⭐ ⭐ Let K be a compact metric space and $\mathscr{C}(K)$ be equicontinuous on K. Prove that $\overline{\mathscr{C}(K)}$ is also equicontinuous on K.

Proof. Since K is compact, every element $f \in \mathscr{C}(K)$ must be bounded by the Extreme Value Theorem [25, p. 100]. Given $\epsilon > 0$. Fix $\theta \in (0, \epsilon)$. Since $\mathscr{C}(K)$ is equicontinuous on K, there exists a $\delta > 0$ such that

$$|f(x) - f(y)| < \frac{\epsilon}{3} \tag{10.58}$$

for all $f \in \mathscr{C}(K)$ and all $x, y \in K$ with $|x - y| < \delta$. Let $F \in \overline{\mathscr{C}(K)}$. Then there is a sequence $\{f_n\} \subseteq \mathscr{C}(K)$ such that

$$\sup_{x \in K} |f_n(x) - F(x)| = \|f_n - F\| \to 0 \tag{10.59}$$

as $n \to \infty$. In other words, there is an $N \in \mathbb{N}$ such that $n \geq N$ implies

$$|f_n(x) - F(x)| < \frac{\epsilon}{3}$$

for all $x \in K$.

Now if $x, y \in K$ and $|x - y| < \delta$, then we follow from the inequality (10.58) and the limit (10.59) that

$$|F(x) - F(y)| \leq |F(x) - f_N(x)| + |f_N(x) - f_N(y)| + |f_N(y) - F(y)| < \frac{\epsilon}{3} + \frac{\epsilon}{3} + \frac{\epsilon}{3} = \epsilon.$$

By Definition 10.10 (Equicontinuity), $\overline{\mathscr{C}(K)}$ is equicontinuous on K, completing the proof of the problem. ∎

10.5   Approximation by Polynomials

Problem 10.33

$(\star)$ *Suppose that every f_i $(i = 1, 2, \ldots, m)$ is a real-valued function on $[a, b]$ such that, for each i, there is a sequence of polynomials converges uniformly to f_i on $[a, b]$. Prove that there also exists a sequence of polynomials converges uniformly to their product $f_1 f_2 \cdots f_m$ on $[a, b]$.*

Proof. Since a polynomial is continuous on $[a, b]$, Theorem 10.6 (Uniform Convergence and Continuity) guarantees that each f_i is also continuous on $[a, b]$. Consequently, the product

$$f = f_1 f_2 \cdots f_m$$

is also continuous on $[a, b]$. Now an application of the Weierstrass Approximation Theorem implies that there exists a sequence of polynomials $\{P_n\}$ defined on $[a, b]$ such that $P_n \to f$ uniformly on $[a, b]$ which completes the proof of the problem. $\blacksquare$

Problem 10.34

$(\star)$ *Suppose that $f, g : [a, b] \to \mathbb{R}$ are continuous. Let $F = \max(f, g)$. Prove that there exists a sequence of $\{P_n\}$ defined on $[a, b]$ such that $P_n \to F$ uniformly on $[a, b]$.*

Proof. Since f, g are continuous on $[a, b]$, $|f - g|$ is also continuous on $[a, b]$. Since we have

$$F = \frac{1}{2}(f + g + |f - g|),$$

F is continuous on $[a, b]$ and our desired result follows directly from the Weierstrass Approximation Theorem. This has completed the proof of the problem. $\blacksquare$

Problem 10.35

$(\star)(\star)$ *Suppose that f is a continuously differentiable function on $[a, b]$. Prove that there exists a sequence of polynomials $\{P_n\}$ such that $P_n \to f$ and $P'_n \to f$ uniformly on $[a, b]$.*

Proof. Since f is continuously differentiable on $[a, b]$, the Weierstrass Approximation Theorem implies the existence of a sequence of polynomials $\{Q_n\}$ such that given $\epsilon > 0$, one can find an $N \in \mathbb{N}$ such that $n \geq N$ implies

$$|Q_n(t) - f'(t)| < \frac{\epsilon}{b - a} \tag{10.60}$$

for all $t \in [a, b]$. Now for each $n \in \mathbb{N}$, we define $P_n : [a, b] \to \mathbb{R}$ by

$$P_n(x) = f(a) + \int_a^x Q_n(t) \, \mathrm{d}t.$$

It is clear that every P_n is a polynomial on $[a, b]$. Furthermore, we deduce from the First and the Second Fundamental Theorem of Calculus (see [25, p. 161]) that

$$f(x) - P_n(x) = [f(x) - f(a)] - [P_n(x) - P_n(a)]$$

$$= \int_a^x f'(t)\,\mathrm{d}t - \int_a^x P_n'(t)\,\mathrm{d}t$$

$$= \int_a^x [f'(t) - Q_n(t)]\,\mathrm{d}t.$$

It follows from the estimate (10.60) that for every $x \in [a, b]$ and $n \geq N$, we have

$$|f(x) - P_n(x)| \leq \int_a^x |f'(t) - Q_n(t)|\,\mathrm{d}t < \frac{\epsilon}{b-a}(x - a) < \epsilon.$$

By Definition 10.2 (Uniform Convergence), $P_n \to f$ uniformly on $[a, b]$. We have completed the proof of the problem.   ∎

Problem 10.36

(⋆)(⋆) *Let $f \in \mathscr{C}([0,1])$ and $\epsilon > 0$. Prove that there exists a polynomial P with rational coefficients on $[0, 1]$ such that*

$$\|f - P\| < \epsilon.$$

Proof. By the Weierstrass Approximation Theorem, we know that there exists a polynomial p on $[0, 1]$ such that

$$\|f - p\| < \frac{\epsilon}{2}. \tag{10.61}$$

Let

$$p(x) = a_0 x^n + a_1 x^{n-1} + \cdots + a_{n-1}x + a_n,$$

where $a_0, a_1, \ldots, a_n$ are constants. For each a_k, there exists a $q_k \in \mathbb{Q}$ such that

$$|a_k - q_k| < \frac{\epsilon}{2(n+1)}.$$

Now we define

$$P(x) = q_0 x^n + q_1 x^{n-1} + \cdots + q_{n-1}x + q_n.$$

Then it is easy to check that for $x \in [0, 1]$, we have

$$|p(x) - P(x)| \leq \sum_{k=0}^{n} |a_k - q_k| \cdot |x^{n-k}| < \sum_{k=0}^{n} \frac{\epsilon}{2(n+1)} = \frac{\epsilon}{2}$$

which means that

$$\|p - P\| \leq \frac{\epsilon}{2}. \tag{10.62}$$

Combining the inequalities (10.61) and (10.62), we see that

$$\|f - P\| \leq \|f - p\| + \|p - P\| < \epsilon.$$

This completes the proof of the problem.   ∎

> ### Problem 10.37
>
> ⋆ ⋆ ⋆ Let $g_{n,m}(x) = C_m^n x^m (1-x)^{n-m}$, where $0 \leq m \leq n$. Define the nth **Bernstein polynomial** of a continuous function $f : [0,1] \to \mathbb{R}$ by
>
> $$B_n(x) = \sum_{m=0}^{n} f\left(\frac{m}{n}\right) g_{n,m}(x).$$
>
> Prove that $B_n \to f$ uniformly on $[0,1]$.

Proof. Recall from [25, Theorem 7.10, p. 100] that f is uniformly continuous on $[0,1]$. In other words, given $\epsilon > 0$, there is a $\delta > 0$ such that

$$|f(x) - f(y)| < \frac{\epsilon}{2} \tag{10.63}$$

for all $x, y \in [0,1]$ with $|x - y| < \delta$. It is clear that

$$1 = [x + (1-x)]^n = \sum_{m=0}^{n} C_m^n x^m (1-x)^{n-m}, \tag{10.64}$$

so we have

$$f(x) - B_n(x) = \sum_{m=0}^{n} \left[f(x) - f\left(\frac{m}{n}\right) \right] g_{n,m}(x)$$
$$= \sum_{1} \left[f(x) - f\left(\frac{m}{n}\right) \right] g_{n,m}(x) + \sum_{2} \left[f(x) - f\left(\frac{m}{n}\right) \right] g_{n,m}(x), \tag{10.65}$$

where the first summation on the right-hand side of (10.65) consists of those values of m for which $|x - \frac{m}{n}| < \delta$ and the second summation takes the rest of the values of m.

Now we are going to find bounds of the two summations in the equation (10.65). To this end, we notice from the inequality (10.63) that

$$\left| \sum_{1} \left[f(x) - f\left(\frac{m}{n}\right) \right] g_{n,m}(x) \right| \leq \sum_{1} \left| f(x) - f\left(\frac{m}{n}\right) \right| \cdot g_{n,m}(x) < \frac{\epsilon}{2} \sum_{1} g_{n,m}(x) \leq \frac{\epsilon}{2}. \tag{10.66}$$

By the Extreme Value Theorem, there exists a $p \in [0,1]$ such that $f(p) = \sup\limits_{x \in [0,1]} |f(x)|$ and this implies that

$$\left| \sum_{2} \left[f(x) - f\left(\frac{m}{n}\right) \right] g_{n,m}(x) \right| \leq \sum_{2} \left| f(x) - f\left(\frac{m}{n}\right) \right| g_{n,m}(x)$$
$$\leq 2M \sum_{2} g_{n,m}(x)$$
$$\leq 2M \sum_{2} \frac{(nx - m)^2}{n^2 \delta^2} g_{n,m}(x)$$
$$\leq \frac{2M}{n^2 \delta^2} \sum_{m=0}^{n} (nx - m)^2 g_{n,m}(x). \tag{10.67}$$

Note that

$$(e^y + 1 - x)^n = \sum_{m=0}^{n} C_m^n e^{my}(1-x)^{n-m}. \tag{10.68}$$

Differentiating the equation (10.68) with respect to y twice, we obtain

$$n(e^y + 1 - x)^{n-1}e^y = \sum_{m=0}^{n} C_m^n m e^{my}(1-x)^{n-m} \tag{10.69}$$

$$n(e^y + 1 - x)^{n-1}e^y + n(n-1)(e^y + 1 - x)^{n-2}e^{2y} = \sum_{m=0}^{n} C_m^n m^2 e^{my}(1-x)^{n-m}. \tag{10.70}$$

By putting $e^y = x$ into the formulas (10.69) and (10.70), they yield that

$$nx = \sum_{m=0}^{n} C_m^n m x^m (1-x)^{n-m} \quad \text{and} \quad nx + n(n-1)x^2 = \sum_{m=0}^{n} C_m^n m^2 x^m (1-x)^{n-m}. \tag{10.71}$$

If we rewrite the identities (10.64) and (10.71) as

$$1 = \sum_{m=0}^{n} g_{n,m}(x), \quad nx = \sum_{m=0}^{n} m g_{n,m}(x) \quad \text{and} \quad nx + n(n-1)x^2 = \sum_{m=0}^{n} m^2 g_{n,m}(x),$$

then they imply that

$$\sum_{m=0}^{n} (nx - m)^2 g_{n,m}(x) = \sum_{m=0}^{n} (n^2 x^2 - 2nmx + m^2) g_{n,m}(x)$$

$$= n^2 x^2 \sum_{m=0}^{n} g_{n,m}(x) - 2nx \sum_{m=0}^{n} m g_{n,m}(x) + \sum_{m=0}^{n} m^2 g_{n,m}(x)$$

$$= n^2 x^2 - 2n^2 x^2 + nx + n(n-1)x^2$$

$$= nx(1-x). \tag{10.72}$$

By substituting the identity (10.72) into the inequality (10.67) and using the fact that the function $h(x) = x(1-x)$ defined on $[0,1]$ attains its absolute maximum $\frac{1}{4}$ at $x = \frac{1}{2}$, we derive

$$\left| \sum_{2} \left[f(x) - f\left(\frac{m}{n}\right) \right] g_{n,m}(x) \right| \le \frac{2M}{n^2 \delta^2} \cdot nx(1-x) \le \frac{M}{2n\delta^2}. \tag{10.73}$$

Recall that δ is *fixed* so that we can choose $n \in \mathbb{N}$ such that $n > \frac{M}{\epsilon \delta^2}$. Then the inequality (10.73) gives

$$\left| \sum_{2} \left[f(x) - f\left(\frac{m}{n}\right) \right] g_{n,m}(x) \right| < \frac{\epsilon}{2}. \tag{10.74}$$

Hence we put the bounds (10.66) and (10.74) back into the expression (10.65) to conclude that

$$|f(x) - B_n(x)| < \epsilon$$

for all $x \in [0,1]$ and $n > \frac{M}{\epsilon \delta^2}$. By Definition 10.2 (Uniform Convergence), $B_n \to f$ uniformly on $[0,1]$ and we complete the analysis of the proof. ∎

CHAPTER **11**

Improper Integrals

11.1 Fundamental Concepts

In [25, Chap. 9], we study an integral whose integrand is a function f **defined** and **bounded** on a **finite** interval $[a, b]$. However, we can extend the concepts of integration to include functions tending to infinity at some certain points or unbounded intervals. This is the content of the so-called **improper integrals**. The main references we have used here are [1, §10.23], [2, §10.13], [16, Chap. 1] and [27, §6.5].

11.1.1 Improper Integrals of the First Kind

Definition 11.1. *Let $a \in \mathbb{R}$. Suppose that $f : [a, +\infty) \to \mathbb{R}$ is a function integrable on every closed and bounded interval $[a, b]$, where $a \leq b < +\infty$, i.e., $f \in \mathscr{R}$ on $[a, b]$. We set*

$$\int_a^{+\infty} f(x) \, \mathrm{d}x = \lim_{b \to +\infty} \int_a^b f(x) \, \mathrm{d}x, \tag{11.1}$$

*where the left-hand side is called the **improper integral of the first kind**. If the limit (11.1) exists and is finite, then we call the improper integral **convergent**. Otherwise, the improper integral is **divergent**.*

Similarly, we may define the following improper integral if the limit exists and is finite:

$$\int_{-\infty}^b f(x) \, \mathrm{d}x = \lim_{a \to -\infty} \int_a^b f(x) \, \mathrm{d}x.$$

Finally, if both

$$\int_{-\infty}^a f(x) \, \mathrm{d}x \quad \text{and} \quad \int_a^{+\infty} f(x) \, \mathrm{d}x$$

are convergent *for some $a \in \mathbb{R}$*, then we have

$$\int_{-\infty}^{+\infty} f(x) \, \mathrm{d}x = \int_{-\infty}^a f(x) \, \mathrm{d}x + \int_a^{+\infty} f(x) \, \mathrm{d}x \tag{11.2}$$

33

so that the improper integral on the left-hand side of the expression (11.2) is convergent. If one of the integrals on the right-hand side of the expression (11.2) is divergent, then the corresponding improper integral is **divergent**.

Remark 11.1

We must notice that it can be shown that the choice of a in the expression (11.2) is *not* important.

11.1.2 Improper Integrals of the Second Kind

Definition 11.2. *Let $a, b \in \mathbb{R}$. Suppose that $f : [a, b) \to \mathbb{R}$ is a function integrable on every closed and bounded interval $[a, c]$, where $a \le c < b$, i.e., $f \in \mathscr{R}$ on $[a, c]$. We set*

$$\int_a^b f(x)\, \mathrm{d}x = \lim_{c \to b-} \int_a^c f(x)\, \mathrm{d}x, \tag{11.3}$$

*where the left-hand side is called the **improper integral of the second kind**. If the limit (11.3) exists and is finite, then we call the improper integral **convergent**. Otherwise, the improper integral is **divergent**.*

The key of the definition is that f may be **unbounded** in a neighborhood of b. Similarly, if we have $f : (a, b] \to \mathbb{R}$, then we define

$$\int_a^b f(x)\, \mathrm{d}x = \lim_{c \to a+} \int_c^b f(x)\, \mathrm{d}x.$$

Definition 11.3. *Let $a, b \in \mathbb{R}$ and $a < c < b$. Suppose that $f : [a, c) \cup (c, b] \to \mathbb{R}$ is a function integrable on every closed and bounded interval of $[a, c) \cup (c, b]$ and $f(x) \to \pm\infty$ as $x \to c$. Then we have*

$$\int_a^b f(x)\, \mathrm{d}x = \int_a^c f(x)\, \mathrm{d}x + \int_c^b f(x)\, \mathrm{d}x \tag{11.4}$$

and the improper integral on the left-hand side of (11.4) exists if and only if the two improper integrals on its right-hand side exist.

Examples of improper integrals of the first and the second kinds can be found in Problems 11.1 and 11.2 respectively.

11.1.3 Cauchy Principal Value

We can define the so-called **Cauchy Principal Value**, or simply **principal value**, of an improper integral in some cases.

Definition 11.4 (Cauchy Principal Value). *Let $a, b \in \mathbb{R}$ and $a < c < b$. We suppose that $f : [a, c) \cup (c, b] \to \mathbb{R}$ is a function integrable on every closed and bounded interval of $[a, c) \cup (c, b]$ and $f(x) \to \pm\infty$ as $x \to c$. Then the principal value of the improper integral of f on $[a, b]$ is given by*

$$\text{P.V.} \int_a^b f(x)\, \mathrm{d}x = \lim_{\epsilon \to 0+} \left(\int_a^{c-\epsilon} f(x)\, \mathrm{d}x + \int_{c+\epsilon}^b f(x)\, \mathrm{d}x \right). \tag{11.5}$$

Now if the improper integral (11.4) exists, then it is easy to see that it equals to its principal value (11.5). However, the converse is not true, see Problem 11.4 for an example.

▌11.1.4 Properties of Improper Integrals

Theorem 11.5. *Suppose that $a \in \mathbb{R}$ and $f, g : [a, b) \to \mathbb{R}$ are integrable on every closed and bounded interval of $[a, b)$, where b is either finite or $+\infty$.*

(a) *For any $A, B \in \mathbb{R}$, the function $Af + Bg$ is also integrable on every closed and bounded interval of $[a, b)$ and furthermore,*

$$\int_a^b (Af + Bg)(x)\,\mathrm{d}x = A \int_a^b f(x)\,\mathrm{d}x + B \int_a^b g(x)\,\mathrm{d}x.$$

(b) *Let $\varphi : [\alpha, \beta) \to [a, b)$ be a strictly increasing differentiable function such that $\varphi(x) \to b$ as $x \to \beta$. Then the function $F : [\alpha, \beta) \to \mathbb{R}$ defined by $F(t) = f(\varphi(t))\varphi'(t)$ satisfies*

$$\int_a^b f(x)\,\mathrm{d}x = \int_\alpha^\beta F(t)\,\mathrm{d}t = \int_\alpha^\beta f(\varphi(t))\varphi'(t)\,\mathrm{d}t.$$

Theorem 11.6. *Suppose that $a \in \mathbb{R}$ and $f : (a, b) \to \mathbb{R}$ is integrable on every closed and bounded interval of (a, b), where b is either finite or $+\infty$. For $a < c < b$, we have*

$$\int_a^b f(x)\,\mathrm{d}x = \int_a^c f(x)\,\mathrm{d}x + \int_c^b f(x)\,\mathrm{d}x. \tag{11.6}$$

Hence the improper integral

$$\int_a^b f(x)\,\mathrm{d}x \tag{11.7}$$

converges if and only if both improper integrals

$$\int_a^c f(x)\,\mathrm{d}x \quad \text{and} \quad \int_c^b f(x)\,\mathrm{d}x$$

converge.

In particular, if $b = +\infty$ in the formula (11.6), then the first and the second improper integrals on its right-hand side are of the first kind and the second kind respectively. In this case, the integral (11.7) is sometimes called the **improper integral of the third kind**. An example of this is the classical **Gamma function** Γ defined by

$$\Gamma(x) = \int_0^{+\infty} t^{x-1}\mathrm{e}^{-t}\,\mathrm{d}t,$$

where $x > 0$. See Problem 11.15 for the proof of its convergence.

▌ 11.1.5 Criteria for Convergence of Improper Integrals

Theorem 11.7 (Comparison Test). *Suppose that $f, g : [a, b) \to [0, +\infty)$, $0 \leq f(x) \leq g(x)$ on $[a, b)$, where b is either finite or $+\infty$, and*

$$\int_a^c f(x)\, dx$$

exists for every $a \leq c < b$. Then we have

$$0 \leq \int_a^b f(x)\, dx \leq \int_a^b g(x)\, dx.$$

Furthermore, the convergence of the improper integral of g implies that of f and the divergence of the improper integral of f implies that of g.

Theorem 11.8 (Limit Comparison Test). *Suppose that $f, g : [a, b) \to [0, +\infty)$ and both*

$$\int_a^c f(x)\, dx \quad \text{and} \quad \int_a^c g(x)\, dx$$

exist for every $a \leq c < b$. Furthermore, if

$$\lim_{x \to b} \frac{f(x)}{g(x)} = A \neq 0, \tag{11.8}$$

where b is either finite or $+\infty$, then the improper integrals

$$\int_a^b f(x)\, dx \quad \text{and} \quad \int_a^b g(x)\, dx$$

both converge or both diverge. If the value of A in the limit (11.8) is 0, then the convergence of $\int_a^b g(x)\, dx$ implies the convergence of $\int_a^b f(x)\, dx$ only.

Theorem 11.9 (Absolute Convergence Test). *Suppose that $f : [a, b) \to \mathbb{R}$ is a function such that*

$$\int_a^c f(x)\, dx$$

exists for every $a \leq c < b$ and the improper integral

$$\int_a^b |f(x)|\, dx$$

converges, where b is either finite or $+\infty$. Then we have

$$\left| \int_a^b f(x)\, dx \right| \leq \int_a^b |f(x)|\, dx$$

and the improper integral

$$\int_a^b f(x)\, dx$$

converges.

Theorem 11.10 (Integral Test for Convergence of Series). *Let N be a positive integer. Suppose that $f : [N, +\infty) \to [0, +\infty)$ decreases on $[N, +\infty)$ and*

$$\int_N^b f(x)\, dx$$

converges for every $N \le b < +\infty$. Then we have

$$\sum_{k=N+1}^{\infty} f(k) \le \int_N^{+\infty} f(x)\, dx \le \sum_{k=N}^{+\infty} f(k).$$

Particularly,

$$\int_N^{+\infty} f(x)\, dx$$

converges if and only if

$$\sum_{k=N}^{+\infty} f(k)$$

converges.

Remark 11.2

Of course, there are analogies of convergence theorems for functions $f : (a, b] \to [0, +\infty)$, where a is either finite or $-\infty$, but we won't repeat the statements here.

11.2 Evaluations of Improper Integrals

Problem 11.1

$\bigstar$ *Evaluate the following improper integrals:*

(a) $\displaystyle \int_1^{+\infty} \frac{dx}{1 + x^2}$.

(b) $\displaystyle \int_1^{+\infty} \frac{dx}{\sqrt{x}}$.

Proof.

(a) By Definition 11.1, we obtain

$$\int_1^{+\infty} \frac{dx}{1 + x^2} = \lim_{b \to +\infty} \int_1^b \frac{dx}{1 + x^2} = \lim_{b \to +\infty} [\arctan x]_1^b = \lim_{b \to +\infty} \left(\arctan b - \frac{\pi}{4} \right) = \frac{\pi}{4}.$$

(b) By Definition 11.1, we have

$$\int_1^{+\infty} \frac{dx}{\sqrt{x}} = \lim_{b \to +\infty} \int_1^b \frac{dx}{\sqrt{x}} = \lim_{b \to +\infty} [2\sqrt{x}]_1^b = \lim_{b \to +\infty} (2\sqrt{b} - 2) = \infty.$$

This completes the proof of the problem.   ■

Problem 11.2

$(\star)$ *Evaluate the following improper integrals:*

(a) $\displaystyle\int_0^1 \frac{\mathrm{d}x}{\sqrt{x}}$.

(b) $\displaystyle\int_1^3 \frac{1}{\sqrt{9-x^2}}\,\mathrm{d}x$.

Proof.

(a) By Definition 11.2, it is easy to check that

$$\int_0^1 \frac{\mathrm{d}x}{\sqrt{x}} = \lim_{\epsilon \to 0+} \int_\epsilon^1 \frac{\mathrm{d}x}{\sqrt{x}} = \lim_{\epsilon \to 0+} [2\sqrt{x}]_\epsilon^1 = \lim_{\epsilon \to 0+} (2 - 2\sqrt{\epsilon}) = 2.$$

(b) By Definition 11.2 and then the substitution $x = 3\sin\theta$, we know that

$$\int_1^3 \frac{1}{\sqrt{9-x^2}}\,\mathrm{d}x = \lim_{c \to 3-} \int_1^c \frac{1}{\sqrt{9-x^2}}\,\mathrm{d}x = \lim_{c \to 3-} \int_0^{\arcsin\frac{c}{3}} \mathrm{d}\theta = \lim_{c \to 3-} \arcsin\frac{c}{3} = \frac{\pi}{2}.$$

Hence we complete the proof of the problem. ∎

Problem 11.3

$(\star)$ *Evaluate the improper integral*

$$\int_1^{+\infty} \frac{1}{x^\alpha}\,\mathrm{d}x,$$

where α is real.

Proof. Clearly, if $b \geq 1$, then we have

$$\int_1^b \frac{1}{x^\alpha}\,\mathrm{d}x = \begin{cases} \left[\dfrac{x^{1-\alpha}}{1-\alpha}\right]_1^b, & \text{if } \alpha \neq 1; \\[2ex] [\ln x]_1^b, & \text{otherwise,} \end{cases}$$

$$= \begin{cases} \dfrac{b^{1-\alpha}-1}{1-\alpha}, & \text{if } \alpha \neq 1; \\[2ex] \ln b, & \text{otherwise.} \end{cases}$$

On the one hand, if $\alpha \neq 1$, then we get from Definition 11.1 that

$$\int_1^{+\infty} \frac{1}{x^\alpha}\,\mathrm{d}x = \lim_{b \to +\infty} \frac{b^{1-\alpha}-1}{1-\alpha} = \begin{cases} \dfrac{1}{\alpha-1}, & \text{if } \alpha > 1; \\[2ex] +\infty, & \text{otherwise.} \end{cases} \tag{11.9}$$

On the other hand, if $\alpha = 1$, then we ahve

$$\int_1^{+\infty} \frac{1}{x}\, dx = \lim_{b \to +\infty} \ln b = +\infty. \tag{11.10}$$

Combining the two results (11.9) and (11.10), we see that

$$\int_1^{+\infty} \frac{1}{x^\alpha}\, dx = \begin{cases} \dfrac{1}{\alpha - 1}, & \text{if } \alpha > 1; \\[2mm] +\infty, & \text{otherwise.} \end{cases}$$

We have completed the proof of the problem.   ∎

Problem 11.4

(⋆) *Evaluate the improper integral of*

$$\int_{-1}^{1} \frac{dx}{x} \tag{11.11}$$

and its principal value.

Proof. We write

$$\int_{-1}^{1} \frac{dx}{x} = \int_{-1}^{0} \frac{dx}{x} + \int_{0}^{1} \frac{dx}{x}. \tag{11.12}$$

Since the two improper integrals on the right-hand side of (11.12) are divergent, the improper integral (11.11) is divergent too. However, we have

$$\text{P.V.} \int_{-1}^{1} \frac{dx}{x} = \lim_{\epsilon \to 0+} \left(\int_{-1}^{-\epsilon} \frac{dx}{x} + \int_{\epsilon}^{1} \frac{dx}{x} \right) = \lim_{\epsilon \to 0+} \left(\ln|x| \Big|_{-1}^{-\epsilon} + \ln|x| \Big|_{\epsilon}^{1} \right) = 0.$$

This completes the proof of the problem.   ∎

Problem 11.5

(⋆) *Let* $-\infty < a < b < +\infty$. *Evaluate the following improper integral*

$$\int_{a}^{b} \frac{dx}{(b - x)^\alpha}$$

if $\alpha < 1$.

Proof. For $a \le c < b$, we have

$$\int_{a}^{c} \frac{dx}{(b - x)^\alpha} = \left[\frac{(b - x)^{1 - \alpha}}{\alpha - 1} \right]_{a}^{c} = \frac{(b - c)^{1 - \alpha} - (b - a)^{1 - \alpha}}{\alpha - 1}.$$

Since $\alpha < 1$, we have

$$\lim_{c \to b-} (b - c)^{1 - \alpha} = 0$$

and then we get from Definition 11.2 that

$$\int_a^b \frac{dx}{(b-x)^\alpha} = \lim_{c \to b^-} \frac{(b-c)^{1-\alpha} - (b-a)^{1-\alpha}}{\alpha - 1} = \frac{(b-a)^{1-\alpha}}{1-\alpha}.$$

We complete the proof of the problem.   ∎

Problem 11.6

(⋆) Prove that

$$\int_0^{+\infty} f\left(ax + \frac{b}{x}\right) dx = \frac{1}{a}\int_0^{+\infty} f(\sqrt{x^2 + 4ab})\, dx,$$

where a and b are positive. (Assume the improper integrals in the question are well-defined.)

Proof. Suppose that

$$y = ax - \frac{b}{x}. \tag{11.13}$$

Then we have

$$ax + \frac{b}{x} = \sqrt{y^2 + 4ab}. \tag{11.14}$$

Adding the two expressions (11.13) and (11.14) to get

$$x = \frac{1}{2a}(y + \sqrt{y^2 + 4ab}) = \varphi(y)$$

so that

$$\frac{dx}{dy} = \varphi'(y) = \frac{1}{2a} \cdot \frac{y + \sqrt{y^2 + 4ab}}{\sqrt{y^2 + 4ab}}.$$

When $x \to 0+$, $y \to -\infty$; when $x \to +\infty$, $y \to +\infty$. By Theorem 11.5(b), we have

$$\int_0^{+\infty} f\left(ax + \frac{b}{x}\right) dx = \frac{1}{2a}\int_{-\infty}^{+\infty} f(\sqrt{y^2 + 4ab}) \cdot \frac{y + \sqrt{y^2 + 4ab}}{\sqrt{y^2 + 4ab}}\, dy$$

$$= \frac{1}{2a}\int_{-\infty}^{0} f(\sqrt{y^2 + 4ab}) \cdot \frac{y + \sqrt{y^2 + 4ab}}{\sqrt{y^2 + 4ab}}\, dy$$

$$+ \frac{1}{2a}\int_0^{+\infty} f(\sqrt{y^2 + 4ab}) \cdot \frac{y + \sqrt{y^2 + 4ab}}{\sqrt{y^2 + 4ab}}\, dy$$

$$= \frac{1}{2a}\int_0^{+\infty} f(\sqrt{y^2 + 4ab}) \cdot \frac{\sqrt{y^2 + 4ab} - y}{\sqrt{y^2 + 4ab}}\, dy$$

$$+ \frac{1}{2a}\int_0^{+\infty} f(\sqrt{y^2 + 4ab}) \cdot \frac{y + \sqrt{y^2 + 4ab}}{\sqrt{y^2 + 4ab}}\, dy$$

$$= \frac{1}{a}\int_0^{+\infty} f(\sqrt{y^2 + 4ab})\, dy.$$

This completes the proof of the problem.   ∎

Problem 11.7

$(\star)$ Suppose that $f : (0,1) \to \mathbb{R}$ is an increasing function and $\int_0^1 f(x)\,dx$ exists. Prove that

$$\int_0^1 f(x)\,dx = \lim_{n \to +\infty} \frac{1}{n} \sum_{k=1}^n f\left(\frac{k}{n}\right).$$

Proof. Let $n \in \mathbb{N}$. Since f is increasing, we see that

$$\int_0^{1-\frac{1}{n}} f(x)\,dx = \sum_{k=1}^{n-1} \int_{\frac{k-1}{n}}^{\frac{k}{n}} f(x)\,dx \leq \sum_{k=1}^{n-1} \int_{\frac{k-1}{n}}^{\frac{k}{n}} f\left(\frac{k}{n}\right)\,dx = \frac{1}{n} \sum_{k=1}^{n-1} f\left(\frac{k}{n}\right)$$

and

$$\int_{\frac{1}{n}}^1 f(x)\,dx = \sum_{k=1}^{n-1} \int_{\frac{k}{n}}^{\frac{k+1}{n}} f(x)\,dx \geq \sum_{k=1}^{n-1} \int_{\frac{k}{n}}^{\frac{k+1}{n}} f\left(\frac{k}{n}\right)\,dx = \frac{1}{n} \sum_{k=1}^{n-1} f\left(\frac{k}{n}\right).$$

Hence we conclude that

$$\int_0^{1-\frac{1}{n}} f(x)\,dx \leq \frac{1}{n} \sum_{k=1}^{n-1} f\left(\frac{k}{n}\right) \leq \int_{\frac{1}{n}}^1 f(x)\,dx. \tag{11.15}$$

Since $\int_0^1 f(x)\,dx$ exists, Definition 11.2 ensures that

$$\int_0^1 f(x)\,dx = \lim_{n \to +\infty} \int_0^{1-\frac{1}{n}} f(x)\,dx = \lim_{n \to +\infty} \int_{\frac{1}{n}}^1 f(x)\,dx. \tag{11.16}$$

Using this fact (11.16) and applying [25, Theorem 5.6, p. 50] to the inequalities (11.15), we have the desired result which completes the proof of the problem. ∎

Problem 11.8

$(\star)(\star)$ Suppose that $0 < \theta < 1$. We define

$$f_\theta(x) = \left[\frac{\theta}{x}\right] - \theta\left[\frac{1}{x}\right]$$

for all $x \in (0,1)$. Evaluate the improper integral

$$\int_0^1 f_\theta(x)\,dx.$$

Proof. We notice that

$$f_\theta(x) = -\left(\frac{\theta}{x} - \left[\frac{\theta}{x}\right]\right) + \theta\left(\frac{1}{x} - \left[\frac{1}{x}\right]\right)$$

which implies that

$$\int_0^1 f_\theta(x)\,dx = -\int_0^1 \left(\frac{\theta}{x} - \left[\frac{\theta}{x}\right]\right) dx + \int_0^1 \theta\left(\frac{1}{x} - \left[\frac{1}{x}\right]\right) dx$$

$$= -\int_0^\theta \left(\frac{\theta}{x} - \left[\frac{\theta}{x}\right]\right) dx - \int_\theta^1 \left(\frac{\theta}{x} - \left[\frac{\theta}{x}\right]\right) dx + \int_0^1 \theta\left(\frac{1}{x} - \left[\frac{1}{x}\right]\right) dx. \qquad (11.17)$$

If $\theta < x \le 1$, then $\left[\frac{\theta}{x}\right] = 0$ so that

$$\int_\theta^1 \left(\frac{\theta}{x} - \left[\frac{\theta}{x}\right]\right) dx = -\int_\theta^1 \frac{\theta}{x}\,dx = -\theta \ln x\Big|_\theta^1 = \theta \ln \theta. \qquad (11.18)$$

Furthermore, the first integral on the right-hand side of (11.17) can be written as

$$\int_0^\theta \left(\frac{\theta}{x} - \left[\frac{\theta}{x}\right]\right) dx = \lim_{\epsilon \to 0+} \int_\epsilon^\theta \left(\frac{\theta}{x} - \left[\frac{\theta}{x}\right]\right) dx. \qquad (11.19)$$

Applying the substitution $x = \theta y$ to the integral on the right-hand side of (11.19), we obtain

$$\int_\epsilon^\theta \left(\frac{\theta}{x} - \left[\frac{\theta}{x}\right]\right) dx = \theta \int_{\epsilon\theta^{-1}}^1 \left(\frac{1}{y} - \left[\frac{1}{y}\right]\right) dy.$$

Therefore, we have

$$\int_0^\theta \left(\frac{\theta}{x} - \left[\frac{\theta}{x}\right]\right) dx = \lim_{\epsilon \to 0+} \theta \int_{\epsilon\theta^{-1}}^1 \left(\frac{1}{y} - \left[\frac{1}{y}\right]\right) dy = \theta \int_0^1 \left(\frac{1}{y} - \left[\frac{1}{y}\right]\right) dy. \qquad (11.20)$$

By putting the expression (11.18) and (11.20) back into the expression (11.17), we establish immediately that

$$\int_0^1 f_\theta(x)\,dx = \theta \ln \theta,$$

completing the analysis of the problem. ∎

11.3   Convergence of Improper Integrals

Problem 11.9

(⋆) Prove that

$$\int_0^{+\infty} e^{-x^2}\,dx$$

is convergent.

Proof. Let $f(x) = e^{-x^2}$ on $[0, +\infty)$ and

$$g(x) = \begin{cases} f(x), & \text{if } x \in [0,1]; \\ e^{-x}, & \text{otherwise.} \end{cases}$$

Since $e^{-x^2} \leq e^{-x}$ for all $x \geq 1$, we have

$$0 \leq f(x) \leq g(x)$$

on $[0, +\infty)$. Since f is bounded on $[0, 1]$, the proper integral

$$\int_0^1 e^{-x^2}\, dx$$

is finite. Furthermore, we have

$$\int_0^{+\infty} g(x)\, dx = \int_0^1 e^{-x^2}\, dx + \int_1^{+\infty} e^{-x}\, dx = \int_0^1 e^{-x^2}\, dx - [e^{-x}]_1^{+\infty} = \int_0^1 e^{-x^2}\, dx + e^{-1}$$

which means that the improper integral

$$\int_0^{+\infty} g(x)\, dx$$

converges. By Theorem 11.7 (Comparison Test), the improper integral

$$\int_0^{+\infty} e^{-x^2}\, dx$$

is convergent. We end the analysis of the proof of the problem. ■

Problem 11.10

(⋆) *Let $p \in \mathbb{R}$. Prove that*

$$\int_1^{+\infty} x^p e^{-x}\, dx$$

is convergent.

Proof. Consider $f(x) = x^p e^{-x}$ and $g(x) = x^{-2}$. Obviously, by repeated use of L'Hôspital's Rule [25, Theorem 8.10, p. 130], we know that

$$\lim_{x \to +\infty} \frac{f(x)}{g(x)} = \lim_{x \to +\infty} \frac{x^{p+2}}{e^x} = 0.$$

Since we have

$$\int_1^{+\infty} x^{-2}\, dx = -[x^{-1}]_1^{+\infty} = 1,$$

Theorem 11.8 (Limit Comparison Test) implies the required result. Hence we have completed the proof of the problem. ■

Problem 11.11

(⋆) *Prove Theorem 11.9 (Absolute Convergence Test).*

Proof. Suppose that $\int_a^b |f(x)|\,\mathrm{d}x$ converges and $\int_a^x f(t)\,\mathrm{d}t$ exists for all $x \in [a,b)$. We know that

$$0 \le f(x) + |f(x)| \le 2|f(x)|,$$

so Theorem 11.7 (Comparison Test) shows that

$$\int_a^b [f(x) + |f(x)|]\,\mathrm{d}x$$

converges. Hence we immediately deduce from Theorem 11.5(a) that the improper integral

$$\int_a^b f(x)\,\mathrm{d}x$$

converges. This completes the proof of the problem.   ■

Problem 11.12

$(\star)$ *Prove that the improper integral*

$$\int_1^{+\infty} \frac{\sin x}{x^2}\,\mathrm{d}x$$

converges.

Proof. We notice that

$$\left| \frac{\sin x}{x^2} \right| \le \frac{1}{x^2}$$

for all $x \in [1, +\infty)$. By Problem 11.3, the improper integral

$$\int_1^{+\infty} \frac{1}{x^2}\,\mathrm{d}x$$

is convergent. By Theorem 11.7 (Comparison Test),

$$\int_1^{+\infty} \left| \frac{\sin x}{x^2} \right|\,\mathrm{d}x$$

is also convergent. Finally, it follows from Theorem 11.9 (Absolute Convergence Test) that the improper integral

$$\int_1^{+\infty} \frac{\sin x}{x^2}\,\mathrm{d}x$$

converges. We have completed the proof of the problem.   ■

Problem 11.13

$(\star)$ *Prove that the improper integral*

$$\int_0^1 (-\ln x)^\alpha\,\mathrm{d}x$$

converges if and only if $\alpha > -1$.

Proof. Let $t = -\ln x$. Then we have $dt = -\frac{1}{x} dx$. When $x \to 0+$, $t \to +\infty$; when $x = 1$, $t = 0$. Thus Theorem 11.5(b) implies that

$$
\begin{aligned}
\int_0^1 (-\ln x)^\alpha \, dx &= -\int_{+\infty}^0 t^\alpha e^{-t} \, dt \\
&= \int_0^{+\infty} t^\alpha e^{-t} \, dt \\
&= \int_0^1 t^\alpha e^{-t} \, dt + \int_1^{+\infty} t^\alpha e^{-t} \, dt.
\end{aligned} \tag{11.21}
$$

Now we are going to investigate the convergence of the two improper integrals of the expression (11.21).

Obviously, we have

$$
\lim_{t \to +\infty} \frac{t^\alpha e^{-t}}{t^{-2}} = \lim_{t \to +\infty} \frac{t^{\alpha+2}}{e^t} = 0
$$

for every $\alpha \in \mathbb{R}$. Therefore, the last integral of the expression (11.21) converges by Theorem 11.8 (Limit Comparison Test). Next, since

$$
\lim_{t \to 0+} \frac{t^\alpha e^{-t}}{t^\alpha} = \lim_{t \to 0+} e^{-t} = 1,
$$

Theorem 11.8 (Limit Comparison Test) again shows that the first integral of the expression (11.21) converges if and only if the improper integral

$$
\int_0^1 t^\alpha \, dt \tag{11.22}
$$

converges.

If $\alpha = -1$, then the integral (11.22) becomes

$$
\int_0^1 \frac{dt}{t} = \lim_{\epsilon \to 0+} \int_\epsilon^1 \frac{dt}{t} = \lim_{\epsilon \to 0+} \ln t \Big|_\epsilon^1 = -\lim_{\epsilon \to 0+} \ln \epsilon
$$

which is definitely divergent. For $\alpha \neq -1$, we have

$$
\int_0^1 t^\alpha \, dt = \lim_{\epsilon \to 0+} \frac{1 - \epsilon^{1+\alpha}}{1 + \alpha}
$$

which is convergent if and only if $\alpha > -1$. Hence the improper integral

$$
\int_0^1 (-\ln x)^\alpha \, dx
$$

converges if and only if $\alpha > -1$, completing the proof of the problem.   ∎

Problem 11.14

$(\star)(\star)$ *Suppose that $f \in \mathscr{R}([0,1])$, f is periodic with period 1 and*

$$\int_0^1 f(x)\,\mathrm{d}x = 0. \tag{11.23}$$

If $\alpha > 0$, prove that the improper integral

$$\int_1^{+\infty} x^{-\alpha} f(x)\,\mathrm{d}x$$

exists. Hint: It is true that

$$\int_1^b x^{-\alpha} f(x)\,\mathrm{d}x = \int_1^b x^{-\alpha}\,\mathrm{d}g,$$

where $b \geq 1$ and $g(x) = \int_1^x f(t)\,\mathrm{d}t$.

Proof. Let $1 \leq x \leq b < +\infty$ and set

$$g(x) = \int_1^x f(t)\,\mathrm{d}t.$$

By the First Fundamental Theorem of Calculus [25, p. 161], g is continuous on $[1,b]$. By the hypothesis (11.23) and the periodicity of f, we get

$$g(x+1) = \int_1^{x+1} f(t)\,\mathrm{d}t = \int_0^x f(t+1)\,\mathrm{d}t = \int_0^1 f(t)\,\mathrm{d}t + \int_1^x f(t)\,\mathrm{d}t = g(x)$$

so that g is also periodic with period 1. Consequently, g is bounded by a positive constant M on $[1, +\infty)$. Next, it follows from the hint that, for every $b \geq 1$, we have

$$\int_1^b x^{-\alpha} f(x)\,\mathrm{d}x = \int_1^b x^{-\alpha}\,\mathrm{d}g$$

$$= [x^{-\alpha} g(x)]_1^b + \alpha \int_1^b x^{-\alpha-1} g(x)\,\mathrm{d}x$$

$$= b^{-\alpha} g(b) - g(1) + \alpha \int_1^b x^{-\alpha-1} g(x)\,\mathrm{d}x. \tag{11.24}$$

Since g is bounded by M on $[1, +\infty)$ and $\alpha > 0$, we have

$$\lim_{b \to +\infty} b^{-\alpha} g(b) = 0. \tag{11.25}$$

In addition, we have

$$|x^{-\alpha-1} g(x)| \leq M x^{-\alpha-1}$$

for all $x \in [1, +\infty)$, so Problem 11.3, Theorems 11.7 (Comparison Test) and 11.9 (Absolute Convergence Test) imply that

$$\int_1^{+\infty} x^{-\alpha-1} g(x)\,\mathrm{d}x \tag{11.26}$$

is convergent for every $\alpha > 0$. Applying the results (11.24) and (11.25) to the expression (11.26) to conclude that

$$\int_1^{+\infty} x^{-\alpha} f(x) \, dx$$

exists, completing the proof of the problem. ∎

Problem 11.15

$(\star)(\star)$ *Prove that*

$$\Gamma(x) = \int_0^{+\infty} t^{x-1} e^{-t} \, dt \tag{11.27}$$

converges for all $x > 0$.

Proof. We write

$$\Gamma(x) = \int_0^1 t^{x-1} e^{-t} \, dt + \int_1^{+\infty} t^{x-1} e^{-t} \, dt. \tag{11.28}$$

Now Problem 11.10 ensures that the second integral of the expression (11.28) converges for every $x \in \mathbb{R}$. For the first integral, we notice that if $a > 0$ and $t = \frac{1}{s}$, then we obtain

$$\int_a^1 t^{x-1} e^{-t} \, dt = \int_1^{\frac{1}{a}} s^{-x-1} e^{-\frac{1}{s}} \, ds.$$

For $s > 1$, since

$$0 \le s^{-x-1} e^{-\frac{1}{s}} \le s^{-x-1}$$

and we know from Problem 11.3 that

$$\int_1^{+\infty} s^{-x-1} \, ds$$

is convergent for all $x > 0$, Theorem 11.7 (Comparison Test) now implies that

$$\int_1^{+\infty} s^{-x-1} e^{-\frac{1}{s}} \, ds$$

converges for all $x > 0$. Hence the first integral of the expression (11.28) and then the improper integral (11.27) both converge for all $x > 0$. This completes the proof of the problem. ∎

Problem 11.16 (Cauchy Criterion for Improper Integrals)

$(\star)(\star)$ *Suppose that $a \in \mathbb{R}$ and $f : [a, b) \to \mathbb{R}$ is integrable on every closed and bounded interval of $[a, b)$, where b is finite or $+\infty$. Then the improper integral*

$$\int_a^b f(x) \, dx \tag{11.29}$$

converges if and only if for every $\epsilon > 0$, there exists a $M \ge a$ such that for all $M \le A, B < b$, we have

$$\left| \int_A^B f(x) \, dx \right| < \epsilon. \tag{11.30}$$

Proof. Suppose that the improper integral (11.29) converges to L. Given $\epsilon > 0$. By Definition 11.1 or 11.2, there exists a $M \geq a$ such that $M \leq c < b$ implies

$$\left| \int_a^c f(x)\,dx - L \right| < \frac{\epsilon}{2}. \tag{11.31}$$

If $M \leq A, B < b$, then the estimate (11.31) gives

$$
\begin{aligned}
\left| \int_A^B f(x)\,dx \right| &= \left| \int_a^B f(x)\,dx - L + L - \int_a^A f(x)\,dx \right| \\
&\leq \left| \int_a^B f(x)\,dx - L \right| + \left| L - \int_a^A f(x)\,dx \right| \\
&< \epsilon.
\end{aligned}
$$

Conversely, we define $F : [a, b) \to \mathbb{R}$ by

$$F(r) = \int_a^r f(x)\,dx.$$

Then our hypothesis implies that F exists on $[a, b)$ and with this notation, the condition (11.30) is equivalent to

$$|F(B) - F(A)| < \epsilon. \tag{11.32}$$

It is easy to see that there exists a $N \in \mathbb{N}$ such that $\frac{1}{N} < b - M$. Therefore, if we write $A_n = b - \frac{1}{n}, B_m = b - \frac{1}{m}, a_n = F(A_n)$ and $a_m = F(B_m)$, then we have $M \leq A_n, B_m < b$ for all $n, m \geq N$ and so the inequality (11.32) means that

$$|a_m - a_n| < \epsilon$$

for all $n, m \geq N$. In other words, $\{a_n\}$ is Cauchy and it converges to the real number L. Hence this fact implies that

$$\int_a^b f(x)\,dx = \lim_{n \to +\infty} \int_a^{b - \frac{1}{n}} f(x)\,dx = \lim_{n \to +\infty} F\left(b - \frac{1}{n}\right) = \lim_{n \to +\infty} a_n = L.$$

This completes the proof of the problem. ∎

Problem 11.17 (Abel's Test for Improper Integrals)

⊛ ⊛ *Suppose that $a \in \mathbb{R}$ and $f : [a, +\infty) \to \mathbb{R}$ is a function which is integrable on every closed and bounded interval of $[a, +\infty)$ and the improper integral $\int_a^{+\infty} f(x)\,dx$ converges. Let $g : [a, +\infty) \to \mathbb{R}$ be a monotonic bounded function. Then the improper integral*

$$\int_a^{+\infty} f(x)g(x)\,dx$$

converges.

Proof. Let $a \le A \le B < +\infty$. Without loss of generality, we may assume that g is increasing on $[A, B]$. Thus the Second Mean Value Theorem for Integrals [7] implies the existence of a $p \in [A, B]$ such that

$$\int_A^B f(x)g(x)\,\mathrm{d}x = g(A)\int_A^p f(x)\,\mathrm{d}x + g(B)\int_p^B f(x)\,\mathrm{d}x. \tag{11.33}$$

By the hypothesis, there exists a positive constant M_1 such that

$$|g(x)| \le M_1$$

on $[a, +\infty)$. Given $\epsilon > 0$. By Problem 11.16 (Cauchy Criterion for Improper Integrals), there is a positive constant $M_2 \ge a$ such that

$$\left|\int_A^p f(x)\,\mathrm{d}x\right| \le \frac{\epsilon}{2M_1} \quad \text{and} \quad \left|\int_p^B f(x)\,\mathrm{d}x\right| \le \frac{\epsilon}{2M_1} \tag{11.34}$$

for all $A \ge M_2 \ge a$. Combining the formula (11.33) and the inequalities (11.34), we see that

$$\left|\int_A^B f(x)g(x)\,\mathrm{d}x\right| \le |g(A)| \cdot \left|\int_A^p f(x)\,\mathrm{d}x\right| + |g(B)| \cdot \left|\int_p^B f(x)\,\mathrm{d}x\right|$$
$$\le M_1 \cdot \frac{\epsilon}{2M_1} + M_1 \cdot \frac{\epsilon}{2M_1}$$
$$= \epsilon$$

for all $A, B \ge M_2 \ge a$. By Problem 11.16 (Cauchy Criterion for Improper Integrals) again, we conclude that the improper integral

$$\int_a^{+\infty} f(x)g(x)\,\mathrm{d}x$$

converges, completing the proof of the problem. ∎

Problem 11.18 (Dirichlet's Test for Improper Integrals)

(★)(★) *Suppose that $a \in \mathbb{R}$ and $f : [a, +\infty) \to \mathbb{R}$ is a function which is integrable on every closed and bounded interval of $[a, +\infty)$ and there exists a positive constant M such that*

$$\left|\int_a^b f(x)\,\mathrm{d}x\right| \le M \tag{11.35}$$

for all $b > a$. Let $g : [a, +\infty) \to \mathbb{R}$ be a monotonic function such that $g(x) \to 0$ as $x \to +\infty$. Then the improper integral

$$\int_a^{+\infty} f(x)g(x)\,\mathrm{d}x$$

converges.

Proof. Given that $\epsilon > 0$. Since $g(x) \to 0$ as $x \to +\infty$, there exists a $M_1 \ge a$ such that

$$|g(x)| < \frac{\epsilon}{4M}$$

for all $x \geq M_1$. Let $A, B \geq M_1 \geq a$. By the Second Mean Value Theorem for Integrals [7] and the hypothesis (11.35), there exists a p between A and B such that

$$\left| \int_A^B f(x)g(x)\,\mathrm{d}x \right| \leq |g(A)| \cdot \left| \int_A^p f(x)\,\mathrm{d}x \right| + |g(B)| \cdot \left| \int_p^B f(x)\,\mathrm{d}x \right|$$

$$\leq |g(A)| \cdot \left(\left| \int_a^p f(x)\,\mathrm{d}x \right| + \left| \int_a^A f(x)\,\mathrm{d}x \right| \right)$$

$$+ |g(B)| \cdot \left(\left| \int_a^B f(x)\,\mathrm{d}x \right| + \left| \int_a^p f(x)\,\mathrm{d}x \right| \right)$$

$$< \frac{\epsilon}{4M} \cdot 2M + \frac{\epsilon}{4M} \cdot 2M$$

$$= \epsilon.$$

Hence Problem 11.16 (Cauchy Criterion for Improper Integrals) shows that the improper integral

$$\int_a^{+\infty} f(x)g(x)\,\mathrm{d}x$$

converges. We have completed the proof of the problem. ■

11.4 Miscellaneous Problems on Improper Integrals

Problem 11.19

(⋆) Suppose that $\int_a^{+\infty} f(x)\,\mathrm{d}x$ converges. Is it true that $f(x) \to 0$ as $x \to +\infty$?

Proof. The answer is negative. We consider the convergence of the improper integral

$$\int_0^{+\infty} \sin x^2\,\mathrm{d}x.$$

By the substitution $t = x^2$, we have

$$\int_0^{+\infty} \sin x^2\,\mathrm{d}x = \int_0^{+\infty} \frac{\sin t}{2\sqrt{t}}\,\mathrm{d}t. \tag{11.36}$$

Since $g(t) = \frac{1}{2\sqrt{t}} \to 0$ as $t \to +\infty$, it follows from Problem 11.18 (Dirichlet's Test for Improper Integrals) that the integral (11.36) converges. However, it is easy to see that the function $f(x) = \sin x^2$ *does not* have limit as $x \to +\infty$ so that $f(x) \to 0$. This completes the proof of the problem. ■

Problem 11.20

⋆ ⋆ *Prove that the improper integral*

$$\int_2^{+\infty} \frac{1}{(\ln x)^{\ln x}}\, \mathrm{d}x \tag{11.37}$$

converges.

Proof. We consider the function $f : [2, +\infty) \to (0, +\infty)$ defined by

$$f(x) = \frac{1}{(\ln x)^{\ln x}}.$$

By differentiation, we know that

$$f'(x) = -\frac{1}{x(\ln x)^{\ln x}}$$

which is negative on $[2, +\infty)$, i.e, f is decreasing on $[2, +\infty)$.

Next, we consider the convergence of the series

$$\sum_{n=2}^{+\infty} \frac{1}{(\ln n)^{\ln n}}. \tag{11.38}$$

Let $a_n = \frac{1}{(\ln n)^{\ln n}}$. Now the decreasing property of f ensures that the sequence $\{a_n\}$ is decreasing. Since

$$\alpha = \limsup_{n \to +\infty} \sqrt[k]{|2^k a_{2^k}|} = \limsup_{n \to +\infty} \sqrt[k]{\frac{2^k}{(\ln 2^k)^{\ln 2^k}}} = \limsup_{n \to +\infty} \frac{2}{(k \ln 2)^{\ln 2}} = 0,$$

we obtain from [25, Theorem 6.7, p. 76] that the series

$$\sum_{k=1}^{+\infty} 2^k \cdot \frac{1}{(\ln 2^k)^{\ln 2^k}}$$

converges. Thus it follows from [18, Theorem 3.27, p. 61] that the series (11.38) is also convergent. Hence Theorem 11.10 (Integral Test for Convergence of Series) ensures that the improper integral (11.37) converges, completing the proof of the problem. ∎

Problem 11.21

⋆ *Suppose that $\int_a^{+\infty} f(x)\, \mathrm{d}x$ converges absolutely and the function g is bounded on $[a, +\infty)$. Prove that the improper integral*

$$\int_a^{+\infty} f(x)g(x)\, \mathrm{d}x \tag{11.39}$$

converges.

Proof. We have $|g(x)| \le M$ on $[a, +\infty)$ for some positive constant M. Since

$$0 \le |f(x)g(x)| \le M|f(x)|$$

on $[a, +\infty)$ and $\int_a^{+\infty} |f(x)|\, dx$ converges, Theorem 11.7 (Comparison Test) shows that

$$\int_a^{+\infty} |f(x)g(x)|\, dx$$

converges. Finally, the convergence of the improper integral (11.39) follows immediately from Theorem 11.9 (Absolute Convergence Test). Hence we have completed the proof of the problem. ∎

Problem 11.22

⋆ ⋆ *Suppose that* $\int_a^{+\infty} f(x)\, dx$ *is convergent and f is monotonic on $[a, +\infty)$. Verify that*

$$\lim_{x \to +\infty} xf(x) = 0.$$

Proof. Without loss of generality, we may assume that f is monotonically decreasing on $[a, +\infty)$. We claim that $f(x) \ge 0$ for all $x \ge a$. Otherwise, there exists a $p \ge a$ such that $f(p) < 0$. By our assumption, for all $x \ge p$, we have $f(x) \le f(p) < 0$ and so

$$\int_p^{+\infty} f(x) \le \int_p^{+\infty} f(p)\, dx = -\infty,$$

a contradiction. Thus this proves the claim.

Given that $\epsilon > 0$. Since $\int_a^{+\infty} f(x)\, dx$ converges, Problem 11.16 (Cauchy Criterion for Improper Integrals) guarantees that there exists a $M \ge a$ such that if $\frac{x}{2} \ge M$, then

$$\left| \int_{\frac{x}{2}}^x f(t)\, dt \right| < \frac{\epsilon}{2}. \tag{11.40}$$

Furthermore, we have

$$\left| \int_{\frac{x}{2}}^x f(t)\, dt \right| = \int_{\frac{x}{2}}^x f(t)\, dt \ge f(x) \cdot \left(x - \frac{x}{2} \right) = \frac{xf(x)}{2}. \tag{11.41}$$

Combining the inequalities (11.40) and (11.41), we conclude that

$$|xf(x)| < \epsilon$$

which is equivalent to saying that $xf(x) \to 0$ as $x \to +\infty$. We complete the proof of the problem. ∎

Problem 11.23 (Frullani's Integral)

$\star$ $\star$ *Suppose that f is continuous on $[0, +\infty)$ and the limit*

$$\lim_{x \to +\infty} f(x)$$

exists. Let $a, b > 0$. Compute the improper integral

$$\int_0^{+\infty} \frac{f(ax) - f(bx)}{x} \, dx$$

Proof. For $0 < r < R < +\infty$, using the method of substitution, we have

$$
\begin{aligned}
\int_r^R \frac{f(ax) - f(bx)}{x} \, dx &= \int_r^R \frac{f(ax)}{x} \, dx - \int_r^R \frac{f(bx)}{x} \, dx \\
&= \int_{ar}^{aR} \frac{f(x)}{x} \, dx - \int_{br}^{bR} \frac{f(x)}{x} \, dx \\
&= \int_{ar}^{br} \frac{f(x)}{x} \, dx - \int_{aR}^{bR} \frac{f(x)}{x} \, dx \\
&= \int_{ar}^{br} f(x) \, d\alpha - \int_{aR}^{bR} f(x) \, d\alpha,
\end{aligned}
\tag{11.42}
$$

where $\alpha(x) = \ln x$. It is clear that α is monotonically increasing on $[ar, br]$ and $[aR, bR]$. Since f is continuous on $[ar, br]$ and $[aR, bR]$, it is bounded on these closed intervals and we follow from [25, Theorem 9.3(a)] that $f \in \mathscr{R}(\alpha)$ on $[ar, br]$ and $[aR, bR]$. Using the form of the First Mean Value Theorem for Integrals [2, Theorem 7.30, p. 160], we see that there exist $p \in [ar, br]$ and $q \in [aR, bR]$ such that

$$
\int_{ar}^{br} f(x) \, d\alpha = f(p) \ln \frac{br}{ar} = f(p) \ln \frac{b}{a} \quad \text{and} \quad \int_{aR}^{bR} f(x) \, d\alpha = f(q) \ln \frac{bR}{aR} = f(q) \ln \frac{b}{a}. \tag{11.43}
$$

If we put the two formulas (11.43) back into the expression (11.42), then we get

$$
\int_r^R \frac{f(ax) - f(bx)}{x} \, dx = [f(p) - f(q)] \ln \frac{b}{a}. \tag{11.44}
$$

As $r \to 0+$ and $R \to +\infty$, we obtain $p \to 0+$ and $q \to +\infty$. Recall that $f(0+) = f(0)$ and $f(+\infty)$ is a real number, so we may take $r \to 0+$ and $R \to +\infty$ in the expression (11.44), we achieve

$$
\int_0^{+\infty} \frac{f(ax) - f(bx)}{x} \, dx = [f(0) - f(+\infty)] \ln \frac{b}{a},
$$

completing the proof of the problem. ∎

Problem 11.24

$(\star)$ *Suppose that* $f : [0, +\infty) \to [0, +\infty)$ *is continuous and*

$$\int_0^{+\infty} f(x)\, dx \tag{11.45}$$

converges. Prove that

$$\lim_{n \to +\infty} \frac{1}{n} \int_0^n x f(x)\, dx = 0.$$

Proof. Suppose that

$$M = \frac{1}{1 + \displaystyle\int_0^{+\infty} f(x)\, dx}.$$

Given $\epsilon > 0$. By the hypothesis (11.45), there is an $N \in \mathbb{N}$ such that $n \geq N$ implies that

$$0 \leq \int_n^{+\infty} f(x)\, dx < \frac{\epsilon}{M}. \tag{11.46}$$

Now for all sufficiently large enough n such that $n > n\epsilon \geq N$, we know from the inequality (11.46) that

$$
\begin{aligned}
0 \leq \frac{1}{n} \int_0^n x f(x)\, dx &= \frac{1}{n} \int_0^{n\epsilon} x f(x)\, dx + \frac{1}{n} \int_{n\epsilon}^n x f(x)\, dx \\
&< \frac{\epsilon}{M} \int_0^{n\epsilon} f(x)\, dx + \int_{n\epsilon}^n f(x)\, dx \\
&< \frac{\epsilon}{M} \int_0^{n\epsilon} f(x)\, dx + \frac{\epsilon}{M} \\
&< \frac{\epsilon}{M} \cdot \left(\int_0^{+\infty} f(x)\, dx + 1 \right) \\
&= \epsilon.
\end{aligned}
$$

This means that

$$\lim_{n \to +\infty} \frac{1}{n} \int_0^n x f(x)\, dx = 0,$$

completing the proof of the problem.   ∎

Problem 11.25

$(\star)(\star)$ *For every* $x > 1$, *prove that*

$$\text{P.V.} \int_0^x \frac{dt}{\ln t}$$

exists.

Proof. Given $\epsilon > 0$ and $x \in [0, 1)$. Now we have

$$\lim_{t \to 0+} \frac{1}{\ln t} = 0,$$

so if we define

$$f(t) = \begin{cases} \dfrac{1}{\ln t}, & \text{if } t > 0; \\ \\ 0, & \text{if } t = 0, \end{cases}$$

then f is well-defined and continuous on $[0, x]$ so that

$$\int_0^x \frac{dt}{\ln t} = \int_0^x f(t)\, dt$$

exists.

When $x > 1$, we notice that

$$\text{P.V.} \int_0^x \frac{dt}{\ln t} = \lim_{\epsilon \to 0+} \left(\int_0^{1-\epsilon} \frac{dt}{\ln t} + \int_{1+\epsilon}^x \frac{dt}{\ln t} \right). \tag{11.47}$$

By Taylor's Theorem [25, p. 131], we can show that

$$\ln t = (t-1) - \frac{(t-1)^2}{2} + \frac{(t-1)^3}{3\xi^3} = (t-1) + \left[\frac{2(t-1)}{3\xi^3} - 1 \right] \cdot \frac{(t-1)^2}{2},$$

where $\xi \in (1, t)$. Let $\alpha(t) = \frac{2(t-1)}{3\xi^3}$. Then it is clear that

$$\lim_{t \to 1} \alpha(t) = 0.$$

Besides, we have the identity

$$\frac{1}{\ln t} = \frac{1}{t-1} - \frac{\alpha(t) - 1}{2 + [\alpha(t) - 1](t-1)} \tag{11.48}$$

so that

$$\lim_{t \to 1} \frac{\alpha(t) - 1}{2 + [\alpha(t) - 1](t-1)} = 0.$$

Therefore, the second term on the right-hand side of the equation (11.48) is continuous and thus integrable around $t = 1$. This fact and the expression (11.48) indicate that the principal value (11.47) depends on the evaluation of the principal value of $\frac{1}{t-1}$ on $[0, x]$ which is

$$\begin{aligned}
\text{P.V.} \int_0^x \frac{dt}{t-1} &= \lim_{\epsilon \to 0+} \left(\int_0^{1-\epsilon} \frac{dt}{t-1} + \int_{1+\epsilon}^x \frac{dt}{t-1} \right) \\
&= \lim_{\epsilon \to 0+} \left(\ln|t-1| \Big|_0^{1-\epsilon} + \ln|t-1| \Big|_{1+\epsilon}^x \right) \\
&= \ln(x-1).
\end{aligned}$$

Hence we conclude that

$$\text{P.V.} \int_0^x \frac{dt}{\ln t}$$

exists, completing the proof of the problem.   ∎

Problem 11.26

$(\star)(\star)$ Suppose that the improper integral $\displaystyle\int_0^{+\infty} f(x)\,dx$ converges. Prove that the improper integral

$$\int_0^{+\infty} e^{-\epsilon x} f(x)\,dx$$

converges for every $\epsilon > 0$.

Proof. Let $1 < A < C$ and fix $\epsilon > 0$. Then $e^{-\epsilon x} \geq 0$ on $[A, C]$. By the Second Mean Value Theorem for Integrals [7], we see that

$$\int_A^C e^{-\epsilon x} f(x)\,dx = e^{-A\epsilon} \int_A^B f(x)\,dx$$

for some $B \in [A, C]$. Thus we have

$$\left| \int_A^C e^{-\epsilon x} f(x)\,dx \right| \leq \left| \int_A^B f(x)\,dx \right|. \tag{11.49}$$

Since $\displaystyle\int_0^{+\infty} f(x)\,dx$ converges, Problem 11.16 (Cauchy Criterion for Improper Integrals) ensures that

$$\left| \int_A^B f(x)\,dx \right| < \epsilon \tag{11.50}$$

for large enough A. Combining the inequalities (11.49) and (11.50), we obtain

$$\left| \int_A^C e^{-\epsilon x} f(x)\,dx \right| < \epsilon$$

for large enough A and our conclusion follows immediately from Problem 11.16 (Cauchy Criterion for Improper Integrals) again. This ends the analysis of the problem. ∎

CHAPTER **12**

Lebesgue Measure

12.1 Fundamental Concepts

In [25, Chap. 9], we recall the basic theory and main results of the Riemann integral of a bounded function f over a bounded and closed interval $[a, b]$ with approximations associated with f and partitions of $[a, b]$. However, the techniques developed so far for Riemann integration is unsatisfactory for general sets. Consequently, only a small class of functions can be "integrable" in the sense of Riemann. This leads to the theories and the notions of the **Lebesgue measure**, **Lebesgue measurable functions** and **Lebesgue integration**. The discussion of these components will be presented in the successive three chapters and for simplicity, we only discuss the Lebesgue theory on the real line. The main references for this chapter are [5, Chap. 1], [9, Chap. 2 - 5], [17, Chap. 2], [21, Chap. 1] and [22, Chap. 7].

12.1.1 Lebesgue Outer Measure

Definition 12.1 (Length of Intervals). *Let I be a nonempty interval of $\mathbb{R}$ and a and b are the end-points of I.[a] We define its length, denoted by $\ell(I)$, to be infinite if I is unbounded. Otherwise, we define*

$$\ell(I) = b - a.$$

Definition 12.2 (Lebesgue Outer Measure). *Let $E \subseteq \mathbb{R}$ and $\{I_k\}$ be a countable collection of nonempty open and bounded intervals covering E, i.e.,*

$$E \subseteq \bigcup_{k=1}^{\infty} I_k.$$

*Then the **Lebesgue outer measure** (or simply **outer measure**) of E, denoted by $m^*(E)$, is given by*

$$m^*(E) = \inf \left\{ \sum_{k=1}^{\infty} \ell(I_k) \,\middle|\, E \subseteq \bigcup_{k=1}^{\infty} I_k \right\},$$

where the infimum takes over all countable collections of nonempty open and bounded intervals covering E.

---

[a] Note that $|a|$ or $|b|$ can be infinite.

Theorem 12.3 (Properties of Outer Measure). *The outer measure m^* satisfies the following properties:*

(a) $m^*(\varnothing) = 0$.

(b) $m^*(I) = \ell(I)$ *for every interval I.*

(c) *(Monotonicity) If $E \subseteq F$, then $m^*(E) \leq m^*(F)$.*

(d) *(Countably Subadditivity) If $\{E_k\}$ is a countable collection of sets of $\mathbb{R}$, then we have*

$$m^*\Big(\bigcup_{n=1}^{\infty} E_k\Big) \leq \sum_{k=1}^{\infty} m^*(E_k).$$

(e) *(Translation invariance) For every $E \subseteq \mathbb{R}$ and $p \in \mathbb{R}$, we have $m^*(p + E) = m^*(E)$, where $p + E = \{p + x \mid x \in E\}$.*

Remark 12.1

The outer measure m^*, however, is *not* **countable additive**. In other words, there exists a countable collection $\{E_k\}$ of disjoint subsets of $\mathbb{R}$ such that

$$m^*\Big(\bigcup_{k=1}^{\infty} E_k\Big) \neq \sum_{k=1}^{\infty} m^*(E_k).$$

12.1.2 Lebesgue Measurable Sets

Let S be any subset of $\mathbb{R}$. If $E \subseteq S$, then we denote $E^c = S \setminus E$.

Definition 12.4 (Lebesgue Measurability). *A set $E \subseteq \mathbb{R}$ is **Lebesgue measurable**, or simply **measurable**, if for every set $A \subseteq \mathbb{R}$, we have*[b]

$$m^*(A) = m^*(A \cap E) + m^*(A \cap E^c).$$

*If E is measurable, then we define the **Lebesgue measure** of E to be*

$$m(E) = m^*(E).$$

If all the sets considered in Theorem 12.3(c) and (d) (Properties of Outer Measure) are measurable, then the corresponding results also hold with m^* replaced by m. That is, if $E \subseteq F$, then

$$m(E) \leq m(F)$$

and if all E_k are measurable, then

$$m\Big(\bigcup_{n=1}^{\infty} E_k\Big) \leq \sum_{n=1}^{\infty} m(E_k). \tag{12.1}$$

Further properties of measurable sets are listed in the following theorem.

---

[b]This definition is due to Carathèodory.

Theorem 12.5 (Properties of Measurable Sets). *Let $E \subseteq \mathbb{R}$. Then we have the following results:*

(a) *If $m^*(E) = 0$, then it is measurable.*

(b) *Every interval I is measurable.*

(c) *If E is measurable, then E^c is also measurable.*

(d) *If E is measurable and $p \in \mathbb{R}$, then $p + E$ is also measurable and $m(p + E) = m(E)$.*

(e) *The finite union or intersection of measurable sets is also measurable.*

> **Remark 12.2**
>
> We notice that any set $E \subseteq \mathbb{R}$ with positive outer measure contains a nonmeasurable subset. See, for examples, [9, Chap. 4, pp. 81 - 83] and [17, Theorem 17, p. 48].

Definition 12.6 (Almost Everywhere). *Let $E \subseteq \mathbb{R}$ be a measurable set. We way that a property P holds **almost everywhere on** E if there exists a subset $A \subseteq E$ such that $m(A) = 0$ and P holds for all $x \in E \setminus A$.*

12.1.3 σ-algebras and Borel Sets

In the following discussion, we suppose that $S \subseteq \mathbb{R}$.

Definition 12.7 (σ-algebra). *Let $\mathfrak{M}$ be a collection of subsets of S. Then $\mathfrak{M}$ is called an σ-algebra in S if $\mathfrak{M}$ satisfies the following conditions:*

(a) $S \in \mathfrak{M}$.

(b) *If $E \in \mathfrak{M}$, then $E^c \in \mathfrak{M}$.*

(c) *If $E = \bigcup_{k=1}^{\infty} E_k$ and if $E_k \in \mathfrak{M}$ for all $k = 1, 2, \ldots$, then $E \in \mathfrak{M}$.*

It follows easily from Definition 12.7 (σ-algebra) that if $E, F, E_k \in \mathfrak{M}$ for all $k = 1, 2, \ldots$, then we have

$$\bigcap_{k=1}^{\infty} E_k^c \in \mathfrak{M} \quad \text{and} \quad E \setminus F \in \mathfrak{M}. \tag{12.2}$$

Theorem 12.8. *The union of a countable collection of measurable sets is also measurable, i.e., the set*

$$E = \bigcup_{k=1}^{\infty} E_k$$

is measurable if all E_k are measurable. Particularly, the collection $\mathfrak{M}$ of all measurable sets of S is an σ-algebra in S.

> **Remark 12.3**
>
> Hence we deduce immediately from the first relation (12.2) and Theorem 12.8 that the intersection of a countable collection of measurable sets is also measurable.

Theorem 12.9. *Every open or closed set in S is measurable.*

Theorem 12.10. *Let $\mathscr{T}$ be a topology of $\mathbb{R}$.*

(a) *The intersection of all σ-algebras in $\mathbb{R}$ containing $\mathscr{T}$ is a smallest σ-algebra $\mathscr{B}$ in $\mathbb{R}$ such that $\mathscr{T} \subseteq \mathscr{B}$.*[c]

(b) *Members of $\mathscr{B}$ are called **Borel sets** of $\mathbb{R}$ and they are **Borel measurable**.*[d]

(c) *The σ-algebra $\mathscr{B}$ is generated by the collection of all open intervals or half-open intervals or closed intervals or open rays or closed rays of $\mathbb{R}$.*[e]

It is well-known that every Borel measurable set is a Lebesgue measurable set, but there exist Lebesgue measurable sets which are *not* Borel measurable, see [5, Exercise 9, p. 48] and [19, Remarks 2.21, p. 53].

▌ 12.1.4 More Properties of Lebesgue Measure

Theorem 12.11 (Countably Additivity)**.** *If $\{E_k\}$ is a countable collection of **disjoint** measurable sets, then $\bigcup_{k=1}^{\infty} E_k$ is measurable and*

$$m\left(\bigcup_{k=1}^{\infty} E_k \right) = \sum_{k=1}^{\infty} m(E_k).$$

Theorem 12.12 (Continuity of Lebesgue Measure)**.** *Suppose that $\{E_k\}$ is a sequence of measurable sets.*

(a) *If $E_k \subseteq E_{k+1}$ for all $k \in \mathbb{N}$, then we have*

$$m\left(\bigcup_{k=1}^{\infty} E_k \right) = \lim_{k \to \infty} m(E_k).$$

(b) *If $E_{k+1} \subseteq E_k$ for all $k \in \mathbb{N}$ and $m(E_1) < \infty$, then we have*

$$m\left(\bigcap_{k=1}^{\infty} E_k \right) = \lim_{k \to \infty} m(E_k).$$

---

[c]The Borel σ-algebra $\mathscr{B}$ is sometimes called the σ-algebra generated by $\mathscr{T}$.
[d]See also Problem 12.35.
[e]In fact, it is [5, Propositin 1.2, p. 22].

12.2 Lebesgue Outer Measure

Problem 12.1

$(\star)$ Prove Theorem 12.5(a).

Proof. Let $m^*(E) = 0$ and A be any set of $\mathbb{R}$. Since $A \cap E \subseteq E$ and $A \cap E^c \subseteq A$, we know from Theorem 12.3(c) (Properties of Outer Measure) that

$$m^*(A \cap E) \leq m^*(E) = 0 \quad \text{and} \quad m^*(A \cap E^c) \leq m^*(A).$$

Therefore, we get

$$m^*(A) \geq m^*(A \cap E) + m^*(A \cap E^c). \tag{12.3}$$

Since $A = (A \cap E) \cup (A \cap E^c)$, we always have

$$m^*(A) \leq m^*(A \cap E) + m^*(A \cap E^c) \tag{12.4}$$

by Theorem 12.3(d) (Properties of Outer Measure). Now the inequalities (12.3) and (12.3) give

$$m^*(A) = m^*(A \cap E) + m^*(A \cap E^c).$$

By Definition 12.4 (Lebesgue Measurability), E is measurable and this completes the proof of the problem. ∎

Problem 12.2

$(\star)$ Let E be open in $\mathbb{R}$. If F and G are subsets of $\mathbb{R}$ such that $F \subseteq E$ and $G \cap E = \varnothing$, prove that

$$m^*(F \cup G) = m^*(F) + m^*(G).$$

Proof. By Theorem 12.9, E is measurable. Since $F \subseteq E$, we have $F \cap E^c = \varnothing$. Since $G \cap E = \varnothing$, we have $G \subseteq E^c$. Hence we see from Definition 12.4 (Lebesgue Measurability) that

$$m^*(F \cup G) = m^*((F \cup G) \cap E) + m^*((F \cup G) \cap E^c) = m^*(F) + m^*(G),$$

as desired. We have completed the proof of the problem. ∎

Problem 12.3

$(\star)$ Construct an unbounded subset of $\mathbb{R}$ with Lebesgue outer measure 0.

Proof. We claim that the set $\mathbb{Q}$ satisfies our requirements. In fact, $\mathbb{Q}$ is clearly unbounded. Furthermore, if $r \in \mathbb{Q}$, then $m^*(\{r\}) = 0$ by Definition 12.2 (Lebesgue Outer Measure). Since we have

$$\mathbb{Q} = \bigcup_{r \in \mathbb{Q}} \{r\},$$

Theorem 12.3(d) (Properties of Outer Measure) implies that

$$m^*(\mathbb{Q}) \le \sum_{r \in \mathbb{Q}} m^*(\{r\}) = 0,$$

completing the proof of the problem. ∎

Problem 12.4

($\star$) *Suppose that E is the set of irrationals in the interval $(0,1)$. Prove that $m^*(E) = 1$.*

Proof. By Theorem 12.3(b) (Properties of Outer Measure), we see that

$$m^*((0,1)) = \ell((0,1)) = 1.$$

Let $F = (0,1) \cap \mathbb{Q}$. By Problem 12.3, we have

$$m^*(F) = 0. \tag{12.5}$$

Since $(0,1) = E \cup F$, it deduces from Theorem 12.3(c) (Properties of Outer Measure) and the result (12.5) that

$$1 = m^*((0,1)) \le m^*(E) + m^*(F) = m^*(E) \le 1$$

so that $m^*(E) = 1$. Hence we have the required result and it completes the proof of the problem. ∎

Problem 12.5

($\star$) *Suppose that $E \subseteq \mathbb{R}$ and $m^*(E) < \infty$. Prove that for every $\epsilon > 0$, there exists an open set $V \subseteq \mathbb{R}$ containing E such that*

$$m^*(V) - m^*(E) < \epsilon. \tag{12.6}$$

Proof. Given $\epsilon > 0$. Then it follows from Definition 12.2 (Lebesgue Outer Measure) that there is a countable collection of intervals $\{I_k\}$ covering E such that

$$\sum_{k=1}^{\infty} \ell(I_k) < m^*(E) + \epsilon. \tag{12.7}$$

Let $V = \bigcup_{k=1}^{\infty} I_k$. We know that V is open in $\mathbb{R}$ and $E \subseteq V$. Using Theorems 12.9, 12.3(b) and (d) (Properties of Outer Measure) and the inequality (12.7), we have

$$m(V) = m^*(V) = m^*\left(\bigcup_{k=1}^{\infty} I_k\right) \le \sum_{k=1}^{\infty} m^*(I_k) = \sum_{k=1}^{\infty} \ell(I_k) < m^*(E) + \epsilon$$

which implies the inequality (12.6). We complete the proof of the problem.

∎

Problem 12.6

$(\star)$ A set $V \subseteq \mathbb{R}$ is called a G_δ set if it is the intersection of a countable collection of open sets of $\mathbb{R}$. Suppose that $E \subseteq \mathbb{R}$ and $m^*(E) < \infty$. Prove that there exists a G_δ set V containing E such that
$$m^*(V) = m^*(E).$$

Proof. By Problem 12.5, for each $k \in \mathbb{N}$, there exists an open set V_k in $\mathbb{R}$ containing E such that
$$m^*(V_k) < m^*(E) + \frac{1}{k}.$$

Let $V = \bigcap_{k=1}^{\infty} V_k$. Then it is clearly a G_δ set by the definition. Furthermore, since $E \subseteq V$, for each $k \in \mathbb{N}$, we must have
$$E \subseteq V \subseteq V_k,$$

so Theorems 12.3(c) (Properties of Outer Measure) and 12.9 imply that
$$m^*(E) \leq m^*(V) \leq m^*(V_k) < m^*(E) + \frac{1}{k}. \tag{12.8}$$

Taking $k \to \infty$ in the inequalities (12.8), we obtain immediately that
$$m^*(V) = m^*(E).$$

This completes the proof of the problem.   ∎

Problem 12.7

$(\star)(\star)$ Let $n \in \mathbb{N}$. Suppose that $\{E_1, E_2, \ldots, E_n\}$ is a finite sequence of disjoint measurable subsets of $\mathbb{R}$. For an arbitrary set $A \subseteq \mathbb{R}$, prove that
$$m^*\Big(A \cap \bigcup_{i=1}^{n} E_i\Big) = \sum_{i=1}^{n} m^*(A \cap E_i). \tag{12.9}$$

Proof. If $n = 1$, then the formula (12.9) is obviously true. Assume that the formula (12.9) is true for $n = k$ for some positive integer k, i.e.,
$$m^*\Big(A \cap \bigcup_{i=1}^{k} E_i\Big) = \sum_{i=1}^{k} m^*(A \cap E_i).$$

Suppose that $\{E_1, E_2, \ldots, E_k, E_{k+1}\}$ is a finite sequence of disjoint measurable sets. Now we follow from the assumption that
$$m^*\Big(A \cap \bigcup_{i=1}^{k} E_i\Big) + m^*(A \cap E_{k+1}) = \sum_{i=1}^{k} m^*(A \cap E_i) + m^*(A \cap E_{k+1}) = \sum_{i=1}^{k+1} m^*(A \cap E_i). \tag{12.10}$$

Since $\displaystyle\bigcup_{i=1}^{k} E_i$ and E_{k+1} are disjoint, we have

$$\left(\bigcup_{i=1}^{k} E_i\right) \cap E_{k+1} = \varnothing \quad \text{and} \quad \left(\bigcup_{i=1}^{k} E_i\right) \cap E_{k+1}^c = \bigcup_{i=1}^{k} E_i$$

which imply that

$$\left(\bigcup_{i=1}^{k+1} E_i\right) \cap E_{k+1} = E_{k+1} \quad \text{and} \quad \left(\bigcup_{i=1}^{k+1} E_i\right) \cap E_{k+1}^c = \bigcup_{i=1}^{k} E_i. \tag{12.11}$$

Substituting the identities (12.11) into the left-hand side of the equation (12.10) and then using the fact that E_{k+1} is measurable, we obtain from Definition 12.4 (Lebesgue Measurability) that

$$m^*\left(A \cap \bigcup_{i=1}^{k} E_i\right) + m^*(A \cap E_{k+1}) = m^*\left(A \cap \left(\left(\bigcup_{i=1}^{k+1} E_i\right) \cap E_{k+1}^c\right)\right)$$

$$+ m^*\left(A \cap \left(\left(\bigcup_{i=1}^{k+1} E_i\right) \cap E_{k+1}\right)\right)$$

$$= m^*\left(\underbrace{\left(A \cap \bigcup_{i=1}^{k+1} E_i\right)}_{\text{The new } A.} \cap E_{k+1}^c\right) + m^*\left(\underbrace{\left(A \cap \bigcup_{i=1}^{k+1} E_i\right)}_{\text{The new } A.} \cap E_{k+1}\right)$$

$$= m^*\left(A \cap \bigcup_{i=1}^{k+1} E_i\right). \tag{12.12}$$

Compare the equations (12.10) and (12.12) to get

$$m^*\left(A \cap \bigcup_{i=1}^{k+1} E_i\right) = \sum_{i=1}^{k+1} m^*(A \cap E_i).$$

By induction, our formula (12.9) is true for every $n \in \mathbb{N}$ and we end the analysis of the problem. ∎

Problem 12.8

$(\star)$ Suppose that $E \subseteq \mathbb{R}$ has positive outer measure. Prove that there exists a bounded subset $F \subseteq E$ such that $m^*(F) > 0$.

Proof. For each $n \in \mathbb{Z}$, we denote $I_n = [n, n+1]$. Then we have

$$E = \bigcup_{n \in \mathbb{Z}} (E \cap I_n).$$

By Theorem 12.3(d) (Properties of Outer Measure), we see that

$$0 < m^*(E) \le \sum_{n \in \mathbb{Z}} m^*(E \cap I_n)$$

which implies that *at least* one of $E \cap I_n$ has positive outer measure. Let $F = E \cap I_N$ be such a set for some $N \in \mathbb{Z}$. Since $F \subseteq I_N$, it must be bounded. Since $F \subseteq E$, this F is the desired set. This completes the proof of the problem.   ∎

12.3 Lebesgue Measurable Sets

Problem 12.9

⊛ *Prove that if E and F are measurable, then*

$$m(E \cup F) + m(E \cap F) = m(E) + m(F).$$

Proof. Notice that
$$E \cup F = (E \setminus F) \cup (E \cap F) \cup (F \setminus E).$$
Obviously, $E \setminus F$, $E \cap F$ and $F \setminus E$ are pairwise disjoint. By Theorem 12.11 (Countably Additive), it is true that

$$m(E \cup F) + m(E \cap F) = [m(E \setminus F) + m(E \cap F)] + [m(F \setminus E) + m(E \cap F)]. \qquad (12.13)$$

Since $E = (E \setminus F) \cup (E \cap F)$ and $F = (F \setminus E) \cup (E \cap F)$, Theorem 12.11 (Countably Additivity) again shows that the equation (12.13) reduces to

$$m(E \cup F) + m(E \cap F) = m(E) + m(F).$$

This completes the proof of the problem.   ∎

Problem 12.10

⊛ *Let $E \subseteq [0,1]$ and $m(E) > 0$. Prove that there exist $x, y \in E$ such that $|x - y|$ is irrational.*

Proof. Assume that $|x - y| \in \mathbb{Q}$ for all $x, y \in E$. Then it means that $E \subseteq x + \mathbb{Q}$ for any $x \in E$. By Problem 12.3 and Theorem 12.3(e) (Properties of Outer Measure), we have

$$m^*(x + \mathbb{Q}) = m^*(\mathbb{Q}) = 0,$$

where $x \in E$. Thus Theorem 12.5(a) (Properties of Measurable Sets) implies that E is measurable and then
$$0 < m(E) = m^*(E) \le m^*(x + \mathbb{Q}) = m(x + \mathbb{Q}) = 0,$$

a contradiction. Hence there exist $x, y \in E$ such that $|x - y|$ is irrational. We end the proof of the problem.   ∎

Problem 12.11

⊛ *Let $E \subseteq [0,1]$ and $m(E) > 0$. Prove that there exist $x, y \in E$ with $x \neq y$ such that $x - y$ is rational.*

Proof. Denote the rationals in $[-1, 1]$ by $r_1, r_2, \ldots$. Define $E_n = r_n + E$ for every $n \in \mathbb{N}$. By Theorem 12.5(d) (Properties of Measurable Sets), we see that

$$m(E_n) = m(r_n + E) = m(E) > 0.$$

Thus we must have

$$\sum_{n=1}^{\infty} m(E_n) = \infty. \tag{12.14}$$

Clearly, we have $E_n \subseteq [-1, 2]$ for every $n \in \mathbb{N}$ so that

$$\bigcup_{n=1}^{\infty} E_n \subseteq [-1, 2]. \tag{12.15}$$

Assume that $E_i \cap E_j = \varnothing$ if $i \neq j$. Then Theorem 12.11 (Countably Additivity) and the set relation (12.15) imply that

$$\sum_{n=1}^{\infty} m(E_n) = m\Big(\bigcup_{n=1}^{\infty} E_n \Big) \leq m([-1, 2]) = 3$$

which contradicts the result (12.14). In other words, there exist $n, m \in \mathbb{N}$ such that

$$E_n \cap E_m \neq \varnothing.$$

Let $z \in E_n \cap E_m$. By the definition, there are $x, y \in E$ such that $z = x + r_n = y + r_m$ which gives

$$x - y = r_m - r_n \in \mathbb{Q}.$$

We have completed the proof of the problem. ∎

Problem 12.12

$(\star)$ *Suppose that $\{r_n\}$ is an enumeration of rationals $\mathbb{Q}$. Prove that*

$$\mathbb{R} \setminus \bigcup_{n=1}^{\infty} \Big(r_n - \frac{1}{n^2}, r_n + \frac{1}{n^2} \Big) \neq \varnothing. \tag{12.16}$$

Proof. Since each $(r_n - \frac{1}{n^2}, r_n + \frac{1}{n^2})$ is measurable, Theorem 12.8 and then Theorem 12.5(c) (Properties of Measurable Sets) imply that the set

$$\mathbb{R} \setminus \bigcup_{n=1}^{\infty} \Big(r_n - \frac{1}{n^2}, r_n + \frac{1}{n^2} \Big)$$

is measurable. By the inequality (12.1), we obtain

$$m\Big(\bigcup_{n=1}^{\infty} \Big(r_n - \frac{1}{n^2}, r_n + \frac{1}{n^2} \Big) \Big) \leq \sum_{n=1}^{\infty} m\Big(\Big(r_n - \frac{1}{n^2}, r_n + \frac{1}{n^2} \Big) \Big) = \sum_{n=1}^{\infty} \frac{2}{n^2} < \infty$$

so that the set relation (12.16) holds. This completes the proof of the problem. ∎

> **Problem 12.13**
>
> $\bigstar$ $\bigstar$ *Suppose that $E \subseteq \mathbb{R}$ and $F \subseteq \mathbb{R}$ are measurable and $E \subseteq F$. Show that $F \setminus E$ is also measurable and*
> $$m(F \setminus E) = m(F) - m(E). \tag{12.17}$$

Proof. We note that

$$F = F \cap \mathbb{R} = F \cap (E \cup E^c) = (F \cap E) \cup (F \cap E^c) = E \cup (F \cap E^c),$$

so we have

$$F \setminus E = F \cap E^c, \tag{12.18}$$

Next, it is clear that

$$E \cap (F \cap E^c) = (E \cap F) \cap (E \cap E^c) = (E \cap F) \cap \varnothing = \varnothing,$$

i.e., E and $F \cap E^c$ are disjoint.

We observe that Definition 12.4 (Lebesgue Measurability) is symmetric in E and E^c, so E^c is measurable and thus the second relation (12.2) guarantees that $F \setminus E$ is also measurable. Since E and $F \cap E^c$ are disjoint, we apply Theorem 12.11 (Countably Additivity) and the set relation (12.18) to obtain

$$m(F) = m(E \cup (F \cap E^c)) = m(E) + m(F \cap E^c) = m(E) + m(F \setminus E)$$

which certainly gives the desired formula (12.17). We complete the proof of the problem. ∎

> **Problem 12.14**
>
> $\bigstar$ *Let N be a positive integer. Suppose that $E_1, E_2, \ldots, E_N \subseteq [0, 1]$ are measurable sets and $\displaystyle\sum_{n=1}^{N} m(E_n) > N - 1$. Prove that*
> $$m\left(\bigcap_{n=1}^{N} E_n \right) > 0.$$

Proof. We have the identity

$$\bigcap_{n=1}^{N} E_n = [0, 1] \setminus \bigcup_{n=1}^{N} E_n',$$

where $E_n' = [0, 1] \setminus E_n$. Since every E_n' $(n = 1, \ldots, N)$ is measurable, the definition and the inequality (12.1) give

$$m\left(\bigcup_{n=1}^{N} E_n' \right) = m\left(\bigcup_{n=1}^{N} ([0, 1] \setminus E_n) \right) \le \sum_{n=1}^{N} m([0, 1] \setminus E_n). \tag{12.19}$$

Furthermore, we apply the identity (12.17) to the right-hand side of the inequality (12.19) to get

$$m\left(\bigcup_{n=1}^{N} E'_n \right) \leq \sum_{n=1}^{N} m([0,1]) - \sum_{n=1}^{N} m(E_n) < N - (N-1) = 1. \tag{12.20}$$

Therefore, using the identity (12.17) and the result (12.20), we establish

$$m\left(\bigcap_{n=1}^{N} E_n \right) = m\left([0,1] \setminus \bigcup_{n=1}^{N} E'_n \right) = m([0,1]) - m\left(\bigcup_{n=1}^{N} E'_n \right) > 1 - 1 = 0.$$

We complete the proof of the problem. ∎

Problem 12.15 (The Borel-Cantelli Lemma)

(⋆) Suppose that $\{E_k\}$ is a countable collection of measurable sets and

$$\sum_{k=1}^{\infty} m(E_k) < \infty.$$

Prove that almost all $x \in \mathbb{R}$ lie in at most finitely many of the sets E_k.

Proof. For each $n \in \mathbb{N}$, let $A_n = \bigcup_{k=n}^{\infty} E_k$. Then we have $A_1 \supseteq A_2 \supseteq \cdots$ and Theorem 12.3(d) (Properties of Outer Measure) implies that

$$m(A_1) = m\left(\bigcup_{k=1}^{\infty} E_k \right) \leq \sum_{k=1}^{\infty} m(E_k) < \infty. \tag{12.21}$$

Apply Theorem 12.12 (Continuity of Lebesgue Measure), we acquire that

$$m\left(\bigcap_{n=1}^{\infty} \left(\bigcup_{k=n}^{\infty} E_k \right) \right) = m\left(\bigcap_{n=1}^{\infty} A_n \right) = \lim_{n \to \infty} m(A_n) \leq \lim_{n \to \infty} \sum_{k=n}^{\infty} m(E_k) = 0.$$

Hence almost all $x \in \mathbb{R}$ fail to lie in the set

$$\bigcap_{n=1}^{\infty} \left(\bigcup_{k=n}^{\infty} E_k \right).$$

In other words, it means that for almost all $x \in \mathbb{R}$ lie in at most finitely many E_k, completing the proof of the problem. ∎

Problem 12.16

(⋆) Let $\{E_k\}$ be a sequence of measurable sets such that

$$m\left(\bigcup_{k=1}^{\infty} E_k \right) < \infty \quad \text{and} \quad \inf_{k \in \mathbb{N}} \{m(E_k)\} = \alpha > 0.$$

Suppose that E is the set of points that lie in an infinity of sets E_k. Prove that E is measurable and $m(E) \geq \alpha$.

Proof. With the same notations as in the proof of Problem 12.15, we see that

$$E = \bigcap_{n=1}^{\infty} A_n$$

and so

$$m(E) = m\Big(\bigcap_{n=1}^{\infty} A_n \Big) = \lim_{n \to \infty} m(A_n). \tag{12.22}$$

Obviously, we have

$$m(A_n) = m\Big(\bigcup_{k=n}^{\infty} E_k \Big) \geq m(E_n) \geq \alpha > 0 \tag{12.23}$$

for every $n \in \mathbb{N}$. Now if we combine the results (12.22) and (12.23), then we get immediately that

$$m(E) \geq \alpha.$$

We complete the proof of the problem. ∎

Problem 12.17

(⋆) *Prove Theorem 12.10(a) and (b).*

Proof. Let Ω be the family of all σ-algebras $\mathfrak{M}$ in $\mathbb{R}$ containing $\mathscr{T}$. It is clear that the collection of all subsets of $\mathbb{R}$ satisfies Definition 12.7 (σ-algebra), so $\Omega \neq \varnothing$. Let

$$\mathscr{B} = \bigcap_{\mathfrak{M} \in \Omega} \mathfrak{M}. \tag{12.24}$$

Since $\mathscr{T} \subseteq \mathfrak{M}$ for all $\mathfrak{M} \in \Omega$, $\mathscr{T} \subseteq \mathscr{B}$. If $E_n \in \mathscr{B}$ for every $n = 1, 2, \ldots$, then the definition (12.24) implies that $E_n \in \mathfrak{M}$ for every $\mathfrak{M} \in \Omega$. Since every $\mathfrak{M}$ is an σ-algebra, we have $\bigcup_{n=1}^{\infty} E_n \in \mathfrak{M}$ for all $\mathfrak{M} \in \Omega$. By the definition (12.24) again, we achieve

$$\bigcup_{n=1}^{\infty} E_n \in \mathscr{B}$$

so that $\mathscr{B}$ satisfies Definition 12.7(c) (σ-algebra). Now we can show the other two parts in a similar way. Hence we conclude that $\mathscr{B}$ is an σ-algebra and it follows from the definition (12.24) that $\mathscr{B}$ is actually a *smallest* σ-algebra containing $\mathscr{T}$.

For the second assertion, by Theorem 12.8, we know that the collection $\mathfrak{M}'$ of all Lebesgue measurable sets of $\mathbb{R}$ is an σ-algebra in $\mathbb{R}$ and by Theorem 12.9, $\mathfrak{M}'$ contains $\mathscr{T}$. In other words, it means that

$$\mathfrak{M}' \in \Omega$$

and thus $\mathscr{B} \subseteq \mathfrak{M}'$ by the definition (12.24). Consequently, a Borel set must be Lebesgue measurable. Hence we have completed the proof of the problem. ∎

Problem 12.18

$(\star)$ *Given $\epsilon > 0$. Construct a dense open set V in $\mathbb{R}$ such that $m(V) < \epsilon$.*

Proof. Let $\{r_n\}$ be a sequence of rational numbers of $\mathbb{R}$. For every n, we consider the open interval I_n given by

$$I_n = (r_n - \epsilon \cdot 2^{-(n+1)}, r_n + \epsilon \cdot 2^{-(n+1)}).$$

Define $V = \bigcup_{n=1}^{\infty} I_n$ which is evidently open in $\mathbb{R}$. Since $\mathbb{Q} \subseteq V$ and $\mathbb{Q}$ is dense in $\mathbb{R}$, V is also dense in $\mathbb{R}$. Finally, we note from Theorem 12.3(d) (Properties of Outer Measure) that

$$m(V) = m\Big(\bigcup_{n=1}^{\infty} I_n \Big) \le \sum_{n=1}^{\infty} m(I_n) = \epsilon \cdot \sum_{n=1}^{\infty} \frac{1}{2^n} = \epsilon,$$

completing the proof of the problem. ∎

Problem 12.19

$(\star)$ *Let C be the Cantor set. Prove that $m(C) = 0$.*

Proof. In [25, Problem 4.28, p. 45], we know that

$$C = \bigcap_{n=1}^{\infty} E_n, \tag{12.25}$$

where

$$E_n = \bigcup_{k=0}^{2^{n-1}-1} \left(\Big[\frac{3k+0}{3^n}, \frac{3k+1}{3^n} \Big] \cup \Big[\frac{3k+2}{3^n}, \frac{3k+3}{3^n} \Big] \right).$$

Since each E_n is the union of closed intervals, E_n is measurable by Theorem 12.8. Thus it follows from Theorem 12.3(d) (Properties of Outer Measure) that

$$
\begin{aligned}
m(E_n) &\le \sum_{k=1}^{2^{n-1}-1} \left\{ m\Big(\Big[\frac{3k+0}{3^n}, \frac{3k+1}{3^n} \Big] \Big) + m\Big(\Big[\frac{3k+2}{3^n}, \frac{3k+3}{3^n} \Big] \Big) \right\} \\
&= \sum_{k=1}^{2^{n-1}-1} \frac{2}{3^n} \\
&= \frac{2(2^{n-1} - 1)}{3^n} \\
&< \Big(\frac{2}{3} \Big)^n.
\end{aligned}
$$

Hence we know from the representation (12.25) that

$$m(C) < \Big(\frac{2}{3} \Big)^n \tag{12.26}$$

for every $n \in \mathbb{N}$. Taking $n \to \infty$ in the inequality (12.26), we conclude that $m(C) = 0$ which completes the analysis of the problem. ∎

Problem 12.20

$(\star)$ *Construct an open subset V of $\mathbb{R}$ such that $m(V) \neq m(\overline{V})$.*

Proof. Let V be the dense open set constructed in Problem 12.18. Then $m(V) < \epsilon$, but $\overline{V} = \mathbb{R}$ which shows that $m(\overline{V}) = \infty$, completing the proof of the problem. ∎

Problem 12.21

$(\star)(\star)$ *Suppose that E and F are subsets of $\mathbb{R}$ with finite outer measure. Prove that*

$$m^*(E \cup F) = m^*(E) + m^*(F) \tag{12.27}$$

if and only if there are measurable sets $E \subseteq E'$ and $F \subseteq F'$ such that

$$m(E' \cap F') = 0. \tag{12.28}$$

Proof. Suppose that there are measurable sets E' and F' with $E \subseteq E'$ and $F \subseteq F'$ such that the equation (12.28) holds. Recall that we always have

$$m^*(E \cup F) \leq m^*(E) + m^*(F).$$

Given $\epsilon > 0$. By Problem 12.5 and Theorem 12.9, there exists an open set V containing $E \cup F$ such that

$$m(V) = m^*(V) < m^*(E \cup F) + \epsilon. \tag{12.29}$$

Since $E \subseteq V \cap E'$, $F \subseteq V \cap F'$ and both $V \cap E'$ and $V \cap F'$ are measurable, we have

$$m^*(E) + m^*(F) \leq m^*(V \cap E') + m^*(V \cap F') = m(V \cap E') + m(V \cap F'). \tag{12.30}$$

By Problem 12.9 and the inequality (12.29), we further reduce the inequality (12.30) to

$$\begin{aligned}
m^*(E) + m^*(F) &\leq m((V \cap E') \cup (V \cap F')) + m((V \cap E') \cap (V \cap F')) \\
&\leq m(V \cap (E' \cup F')) + m(E' \cap F') \\
&\leq m(V) + 0 \\
&< m^*(E \cup F) + \epsilon.
\end{aligned}$$

Since ϵ is arbitrary, we obtain $m^*(E) + m^*(F) \leq m^*(E \cup F)$ which implies the equation (12.27).

Conversely, we suppose that the equation (12.27) holds. By Problem 12.6, there are G_δ sets E' and F' containing E and F respectively such that

$$m(E') = m^*(E') = m^*(E) \quad \text{and} \quad m(F') = m^*(F') = m^*(F).$$

Assume that $m(E' \cap F') > 0$. Then Problem 12.9 implies that

$$\begin{aligned}
m^*(E \cup F) &= m^*(E) + m^*(F) \\
&= m(E') + m(F') \\
&= m(E' \cup F') + m(E' \cap F')
\end{aligned}$$

$$> m(E' \cup F')$$
$$\geq m^*(E' \cup F')$$

which is a contradiction. Consequently, we obtain $m(E' \cap F') = 0$. This completes the proof of the problem. ∎

Problem 12.22

$(\star)$ *Prove that $\mathbb{Q}$ and $\mathbb{R} \setminus \mathbb{Q}$ are Borel sets in $\mathbb{R}$.*

Proof. Let $q \in \mathbb{Q}$. Since $\mathscr{T} \subseteq \mathscr{B}$ and $\mathscr{B}$ is an σ-algebra, Definition 12.7 (σ-algebra) shows that $\mathscr{B}$ also contains all closed subsets of $\mathbb{R}$. Since $\{q\}$ is closed in $\mathbb{R}$, we have $\{q\} \in \mathscr{B}$ and the countability of $\mathbb{Q}$ implies that

$$\mathbb{Q} \in \mathscr{B}.$$

By Definition 12.7 (σ-algebra) again, we have

$$\mathbb{R} \setminus \mathbb{Q} \in \mathscr{B}$$

and hence we complete the proof of the problem. ∎

Problem 12.23

$(\star)(\star)$ *Define*
$$\mathfrak{M} = \{E \subseteq \mathbb{R} \,|\, E \text{ or } E^c \text{ is at most countable}\}.$$

Prove that $\mathfrak{M}$ is an σ-algebra in $\mathbb{R}$.

Proof. We check Definition 12.7 (σ-algebra). Since $\mathbb{R}^c = \varnothing$, we have $\mathbb{R} \in \mathfrak{M}$. Let $E \in \mathfrak{M}$. If E^c is at most countable, then $E^c \in \mathfrak{M}$. Otherwise, E is at most countable. In this case, since $(E^c)^c = E$ is at most countable, we still have $E^c \in \mathfrak{M}$. Suppose that $E_k \in \mathfrak{M}$ for all $k = 1, 2, \ldots$. If *all* E_k are at most countable, then the set

$$E = \bigcup_{k=1}^{\infty} E_k$$

must be at most countable which means that $E \in \mathfrak{M}$. Next, suppose that *there exists* an uncountable E_k. Without loss of generality, we may assume that it is E_1. Since $E_1 \in \mathfrak{M}$, E_1^c is at most countable. Since

$$E^c = \bigcap_{k=1}^{\infty} E_k^c \subseteq E_1^c,$$

E^c must be at most countable. In other words, $E \in \mathfrak{M}$ and we follow from Definition 12.7 (σ-algebra) that $\mathfrak{M}$ is an σ-algebra. This ends the proof of the problem. ∎

Problem 12.24

$(\star)(\star)$ *Suppose that $E \subset \mathbb{R}$ is compact and $K_n = \{x \in \mathbb{R} \,|\, d(x, E) < \frac{1}{n}\}$ for all $n \in \mathbb{N}$, where $d(x, E) = \inf\{d(x, y) \,|\, y \in E\}$. Prove that*

$$m(E) = \lim_{n \to \infty} m(K_n).$$

Proof. It is clear that

$$E \subseteq \bigcap_{n=1}^{\infty} K_n.$$

Let $x \in \bigcap_{n=1}^{\infty} K_n \setminus E \subseteq E^c$. Since E is compact, it is closed in $\mathbb{R}$ and then E^c is open in $\mathbb{R}$. By the definition, there is a $\epsilon > 0$ such that $N_\epsilon(x) \subseteq E^c$ and this means that

$$d(x, E) \geq \epsilon. \tag{12.31}$$

Pick an $N \in \mathbb{N}$ such that $N > \frac{1}{\epsilon}$. Recall that $x \in \bigcap_{n=1}^{\infty} K_n$, so $x \in K_N$. However, the definition of K_N shows that

$$d(x, E) < \frac{1}{N} < \epsilon$$

which contradicts the inequality (12.31). Therefore, no such x exist and it is equivalent to

$$E = \bigcap_{n=1}^{\infty} K_n.$$

Next, since E is bounded, every K_n is also bounded. In particular, we have $m(K_1) < \infty$. It is obvious that

$$K_{n+1} \subseteq K_n$$

for every $n \in \mathbb{N}$. Hence we deduce from Theorem 12.12 (Continuity of Lebesgue Measure), we conclude that

$$m(E) = m\left(\bigcap_{n=1}^{\infty} K_n \right) = \lim_{n \to \infty} m(K_n),$$

completing the proof of the problem. ∎

Problem 12.25

$(\star)(\star)$ *Suppose that $E_1, E_2, \ldots$ are measurable subsets of $\mathbb{R}$ and $m(E_i \cap E_j) = 0$ for all $i \neq j$. Let $E = \bigcup_{n=1}^{\infty} E_n$. Prove that*

$$m(E) = \sum_{n=1}^{\infty} m(E_n).$$

Proof. Define $F_1 = E_1, F_2 = E_2 \setminus E_1, F_3 = E_3 \setminus (E_1 \cup E_2), \ldots$ and in general

$$F_n = E_n \setminus \bigcup_{k=1}^{n-1} E_k \tag{12.32}$$

for every $n = 2, 3, \ldots$. By Theorem 12.8, the collection $\mathfrak{M}$ of all measurable sets of $\mathbb{R}$ is an σ-algebra in $\mathbb{R}$. Recall from the proof of Problem 12.13 that

$$A \setminus B = A \cap B^c,$$

so the representation (12.32) can be rewritten as

$$F_n = E_n \cap \Big(\bigcup_{k=1}^{n-1} E_k \Big)^c = E_n \cap \bigcap_{k=1}^{n-1} E_k^c$$

for each $n = 2, 3, \ldots$. Since $E_n, E_1^c, E_2^c, \ldots, E_{n-1}^c \in \mathfrak{M}$, we see that $F_n \in \mathfrak{M}$ for each $n = 2, 3, \ldots$.

Clearly, we have $F_i \cap F_j = \varnothing$ for $i \neq j$ and

$$F_1 \cup F_2 \cup \cdots \cup F_n = E_1 \cup E_2 \cup \cdots \cup E_n$$

for all $n \in \mathbb{N}$. As a result, we have

$$E = \bigcup_{n=1}^{\infty} E_n = \bigcup_{n=1}^{\infty} F_n.$$

Since each $E_n \in \mathfrak{M}$, the set

$$E_n \cap \Big(\bigcup_{k=1}^{n-1} E_k \Big)$$

is also an element of $\mathfrak{M}$. Thus it follows from Theorem 12.11 (Countably Additivity) that

$$m(E) = m\Big(\bigcup_{n=1}^{\infty} F_n \Big) = \sum_{n=1}^{\infty} m(F_n) \tag{12.33}$$

and

$$m(E_n) = m\Big(E_n \cap \Big(\bigcup_{k=1}^{n-1} E_k \Big) \Big) + m\Big(E_n \cap \Big(\bigcup_{k=1}^{n-1} E_k \Big)^c \Big)$$

$$= m\Big(E_n \cap \Big(\bigcup_{k=1}^{n-1} E_k \Big) \Big) + m(F_n). \tag{12.34}$$

Furthermore, since

$$E_n \cap \Big(\bigcup_{k=1}^{n-1} E_k \Big) = \bigcup_{k=1}^{n-1} (E_n \cap E_k),$$

Theorem 12.3(d) (Properties of Outer Measure) and our hypothesis give

$$m\Big(E_n \cap \Big(\bigcup_{k=1}^{n-1} E_k \Big) \Big) \leq \sum_{k=1}^{n-1} m(E_n \cap E_k) = 0. \tag{12.35}$$

Combining the equation (12.34) and the result (12.35), we conclude immediately that

$$m(F_n) = m(E_n)$$

for every $n = 2, 3, \ldots$. It is trivial that $m(F_1) = m(E_1)$. Hence our desired result derives from these and the formula (12.33). This completes the proof of the problem. ∎

> **Problem 12.26**
>
> ⋆ ⋆ ⋆ *Suppose that E is Lebesgue measurable and $m(E) < \infty$. Prove that there exists a $p \in \mathbb{R}$ such that*
> $$m(E \cap (-\infty, p)) = m(E \cap (p, \infty)).$$

Proof. Define $f : \mathbb{R} \to [0, \infty)$ by

$$f(x) = m(E \cap (-\infty, x)).$$

Now f is an increasing function. Let $x < y$. Since E and (x, y) are measurable, their intersection $E \cap (x, y)$ is also measurable. Since $(x, y) = (-\infty, y) \setminus (-\infty, x]$, we have

$$E \cap (x, y) = E \cap [(-\infty, y) \setminus (-\infty, x]] = [E \cap (-\infty, y)] \setminus [E \cap (-\infty, x]].$$

Obviously, $E \cap (-\infty, x] \subseteq E \cap (-\infty, y)$, so Problem 12.13 implies that

$$
\begin{aligned}
m(E \cap (x, y)) &= m(E \cap (-\infty, y)) - m(E \cap (-\infty, x]) \\
&= m(E \cap (-\infty, y)) - m(E \cap (-\infty, x)) \\
&= f(y) - f(x).
\end{aligned}
$$

By Theorem 12.3(c) (Properties of Outer Measure), we see that

$$0 \le f(y) - f(x) \le m((x, y)) = y - x$$

which implies that

$$|f(x) - f(y)| \le |x - y|$$

for every $x, y \in \mathbb{R}$. In other words, f is a continuous function in $\mathbb{R}$. Since

$$\lim_{x \to -\infty} f(x) = 0 \quad \text{and} \quad \lim_{x \to \infty} f(x) = m(E),$$

the Intermediate Value Theorem [25, p. 101] ensures that there exists a $p \in \mathbb{R}$ such that

$$\frac{1}{2} m(E) = f(p) = m(E \cap (-\infty, p)). \tag{12.36}$$

We note that $E = [E \cap (-\infty, p)] \cup [E \cap [p, \infty)]$ and $[E \cap (-\infty, p)] \cap [E \cap [p, \infty)] = \varnothing$, so we have

$$m(E) = m(E \cap (-\infty, p)) + m(E \cap [p, \infty))$$

and then

$$m(E \cap (p, \infty)) = m(E \cap [p, \infty)) = \frac{1}{2} m(E). \tag{12.37}$$

Therefore, the two expressions (12.36) and (12.37) give the desired result which completes the analysis of the problem. ∎

Problem 12.27

$\star$ $\star$ *Suppose that E is a bounded measurable set with $m(E) > 0$. Let $0 < \alpha < m(E)$. Prove that there exists a measurable set $F \subset E$ such that $m(F) = \alpha$.*

Proof. Since E is bounded, there exist real numbers a and b with $a < b$ such that $E \subseteq [a, b]$. Define $f : [a, b] \to [0, \infty)$ by

$$f(x) = m([a, x] \cap E).$$

Using similar idea as in the proof of Problem 12.26, it can be shown easily that f is an increasing continuous function. Furthermore, we know that

$$f(a) = 0 \quad \text{and} \quad f(b) = m(E),$$

so the Intermediate Value Theorem implies that there is a $p \in [a, b]$ such that

$$f(p) = m([a, p] \cap E) = \alpha.$$

Since $[a, p]$ is measurable, the set $F = [a, p] \cap E \subset E$ is also measurable. This completes the proof of the problem. ∎

Problem 12.28

$\star$ $\star$ *Prove that for every $\epsilon > 0$, there exists an open set $W \subseteq [0, 1]$ such that*

$$\mathbb{Q} \cap [0, 1] \subset W \quad \text{and} \quad m(W) < \epsilon.$$

Then construct a closed subset K of $[0, 1]$ such that

$$m(K) > 0 \quad \text{and} \quad K^\circ = \varnothing.$$

Proof. By Problem 12.18, there exists an open set V containing $\mathbb{Q}$ in $\mathbb{R}$ such that $m(V) < \epsilon$. If we consider $W = V \cap [0, 1]$, then it is open in $[0, 1]$ and it contains $\mathbb{Q} \cap [0, 1]$. It is clear that

$$m(W) \leq m(V) < \epsilon. \tag{12.38}$$

This prove the first assertion.

For the second assertion, we suppose that $K = [0, 1] \setminus W$. Then K is certainly closed in $[0, 1]$. By the estimate (12.38) and using Problem 12.13, we see that

$$m(K) = m([0, 1] \setminus W) = m([0, 1]) - m(W) \geq 1 - \epsilon > 0.$$

Assume that $K^\circ \neq \varnothing$. Then we pick $p \in K^\circ$. Since K° is open in $[0, 1]$ by [25, Problem 4.11, pp. 35, 36], there exists a $\delta > 0$ such that

$$N_\delta(p) \subseteq K^\circ \subseteq K \subseteq [0, 1].$$

However, the neighborhood $N_\delta(p)$ must contain a rational r of $[0, 1]$ which contradicts the fact that K contains no rationals in $[0, 1]$. Hence we complete the proof of the problem. ∎

Remark 12.4

By the second assertion of Problem 12.28, we see that the concept of Lebesgue measure zero is different from empty interior.

Problem 12.29

$(\star)(\star)$ *Suppose that $\{a_n\}$ is a sequence of real numbers and $\{\alpha_n\}$ is a sequence of positive numbers such that $\sum\limits_{n=1}^{\infty} \sqrt{\alpha_n} < \infty$. Prove that there corresponds a measurable set E such that*

$$m(E^c) = 0 \quad \text{and} \quad \sum_{n=1}^{\infty} \frac{\alpha_n}{|x - a_n|} < \infty \tag{12.39}$$

for all $x \in E$.

Proof. Consider the following set

$$E = \{x \in \mathbb{R} \,|\, \text{there is an } N \in \mathbb{N} \text{ such that } n \geq N \text{ implies } |x - a_n| \geq \sqrt{\alpha_n}\}. \tag{12.40}$$

If $x \in E$, then we have

$$\frac{\alpha_n}{|x - a_n|} \leq \sqrt{\alpha_n}$$

for all $n \geq N$. By the hypothesis, we know that

$$\sum_{n=1}^{\infty} \frac{\alpha_n}{|x - a_n|} \leq \sum_{n=1}^{\infty} \sqrt{\alpha_n} < \infty$$

so that the convergence of the infinite series (12.39) is satisfied. Thus it remains to prove that

$$m(E^c) = 0.$$

To see this, we consider the intervals

$$I_n = (a_n - \sqrt{\alpha_n}, a_n + \sqrt{\alpha_n}),$$

where $n = 1, 2, \ldots$. On the one hand, we notice that

$$\sum_{n=1}^{\infty} m(I_n) = \sum_{n=1}^{\infty} 2\sqrt{\alpha_n} < \infty,$$

so $\{I_n\}$ satisfies the hypothesis of Problem 12.15 (The Borel-Cantelli Lemma). Thus almost all $x \in \mathbb{R}$ lie in *at most* finitely many of the sets I_n. On the other hand, $x \in E^c$ if and only if

$$|x - a_n| < \sqrt{\alpha_n}$$

for infinitely many n by the definition (12.40) if and only if $x \in I_n$ for infinitely many n. Consequently, we conclude that $m(E^c) = 0$ which completes the proof of the problem. ∎

12.4 Necessary and Sufficient Conditions for Measurable Sets

Problem 12.30

$(\star)$ *Suppose that E and F differ by a set of measure 0. Prove that E is measurable if and only if F is measurable.*

Proof. By the hypothesis, we have

$$m((E \setminus F) \cup (F \setminus E)) = 0$$

so that $m(E \setminus F) = 0$. Thus it yields from this and Theorem 12.3(d) (Properties of Outer Measure) that

$$
\begin{aligned}
m^*(E \cup F) = m^*((E \setminus F) \cup F) \\
\leq m^*(E \setminus F) + m^*(F) \\
= m(E \setminus F) + m^*(F) \\
= m^*(F) \\
\leq m^*(E \cup F).
\end{aligned}
$$

In other words, it means that $m^*(E \cup F) = m^*(F)$. Similarly, we have $m^*(E \cup F) = m^*(E)$. These two relations imply that

$$m^*(E) = m^*(F).$$

Hence it follows from Definition 12.4 (Lebesgue Measurability) that E is measurable if and only if F is measurable. This ends the proof of the problem. ∎

Problem 12.31

$(\star)(\star)$ *The set $E \subseteq \mathbb{R}$ is measurable if and only if for every $\epsilon > 0$, there exists an open set $V \subseteq \mathbb{R}$ containing E such that*
$$m^*(V \setminus E) < \epsilon. \tag{12.41}$$

Proof. Suppose that E is measurable. Given $\epsilon > 0$. We consider the case that $m(E) < \infty$. Since $m^*(E) = m(E) < \infty$, it follows from Problem 12.5 that there exists an open set $V \subseteq \mathbb{R}$ such that

$$m(V) - m(E) < \epsilon. \tag{12.42}$$

By Problem 12.13, we have

$$m(V \setminus E) = m(V) - m(E). \tag{12.43}$$

Combining the inequality (12.42) and the result (12.43), we establish the expected result (12.41).

Next, we consider the case that $m^*(E) = \infty$. Now, for each $k \in \mathbb{N}$, we define

$$E_k = E \cap [k, k+1).$$

By Theorem 12.3(c) (Properties of Outer Measure), we observe that

$$m(E_k) \le m^*([k, k+1)) = 1$$

so that the result in the previous paragraph can be applied to each E_k. In other words, for each positive integer k, there is an open set V_k containing E_k such that

$$m(V_k \setminus E_k) < \frac{\epsilon}{2^k}.$$

It is clear that the set $V = \bigcup_{k=1}^{\infty} V_k$ is open in $\mathbb{R}$ containing E. Furthermore, we have

$$V \setminus E \subseteq \bigcup_{k=1}^{\infty} (V_k \setminus E_k)$$

and this implies that

$$m(V \setminus E) \le \sum_{k=1}^{\infty} m(V_k \setminus E_k) < \sum_{k=1}^{\infty} \frac{\epsilon}{2^k} = \epsilon$$

which is the case (12.41) again.

Conversely, we suppose that the hypothesis (12.41) holds. Thus for each $k \in \mathbb{N}$, we can choose an open set V_k containing E such that

$$m^*(V_k \setminus E) < \frac{\epsilon}{2^k}. \tag{12.44}$$

Now if we define $V = \bigcup_{k=1}^{\infty} V_k$, then it is an open set in $\mathbb{R}$ and we have

$$V \setminus E = \bigcup_{k=1}^{\infty} (V_k \setminus E)$$

so that we may apply Theorem 12.3(d) (Properties of Outer Measure) to the inequality (12.44) to get

$$m^*(V \setminus E) \le \sum_{k=1}^{\infty} m^*(V_k \setminus E) < \epsilon.$$

Since ϵ is arbitrary, we have $m^*(V \setminus E) = 0$ and Theorem 12.5(a) (Properties of Measurable Sets) ensures that $V \setminus E$ is measurable. Finally, we use the fact $E = V \setminus (V \setminus E)$ and the set relation (12.2) to conclude that E is measurable and hence we complete the proof of the problem. ■

Remark 12.5

Similar to the proof of Problem 12.31, we can show that E is measurable if and only if for every $\epsilon > 0$, there exists a closed set K in $\mathbb{R}$ contained in E such that

$$m^*(E \setminus K) < \epsilon. \tag{12.45}$$

Problem 12.32

$(\star)(\star)$ *Let $E \subseteq \mathbb{R}$. Prove that E is measurable if and only if for every $\epsilon > 0$, there exist an open set V and a closed set K such that*

$$K \subseteq E \subseteq V \quad \text{and} \quad m(V \setminus K) < \epsilon. \tag{12.46}$$

Proof. Suppose that E is measurable. Then we know from the inequalities (12.41) and (12.45) that there exist an open set V and a closed set K such that

$$K \subseteq E \subseteq V, \quad m^*(V \setminus E) < \frac{\epsilon}{2} \quad \text{and} \quad m^*(E \setminus K) < \frac{\epsilon}{2}. \tag{12.47}$$

Since V and K are measurable by Theorem 12.9, both $V \setminus E$ and $E \setminus K$ are measurable and the inequalities (12.47) are also true with m^* replaced by m. Since

$$V \setminus K = (V \setminus E) \cup (E \setminus K),$$

it follows from Theorem 12.3(d) (Properties of Outer Measure) that

$$m(V \setminus K) = m((V \setminus E) \cup (E \setminus K)) \le m(V \setminus E) + +m(E \setminus K) < \epsilon.$$

Conversely, we suppose that the hypotheses (12.46) hold. Since $V \setminus E \subseteq V \setminus K$, it deduces from Theorem 12.3(c) (Properties of Outer Measure) that

$$m^*(V \setminus E) \le m^*(V \setminus K) = m(V \setminus K) < \epsilon.$$

Since ϵ is arbitrary, we actually have $m^*(V \setminus E) = 0$. By Theorem 12.5(a) (Properties of Measurable Sets), we see that $V \setminus E$ is measurable. As $E = V \setminus (V \setminus E)$, we conclude that E is measurable. We complete the proof of the problem. ∎

Problem 12.33

$(\star)(\star)$ *The set $E \subseteq \mathbb{R}$ is measurable if and only if there exists a G_δ set $U \subseteq \mathbb{R}$ containing E such that*

$$m^*(U \setminus E) = 0. \tag{12.48}$$

Proof. Suppose that E is measurable. By Problem 12.31, for each positive integer k, we can select an open set U_k containing E and

$$m^*(U_k \setminus E) < \frac{1}{k}.$$

We define

$$U = \bigcap_{k=1}^{\infty} U_k.$$

By the definition, U is a G_δ set containing E. In addition, for each $k \in \mathbb{N}$, we have $U \setminus E \subseteq U_k \setminus E$, so Theorem 12.3(c) (Properties of Outer Measure) shows that

$$m^*(U \setminus E) \le m^*(U_k \setminus E) < \frac{1}{k}. \tag{12.49}$$

Taking $k \to \infty$ in the inequality (12.49), we conclude that $m^*(U \setminus E) = 0$.

Conversely, we suppose that the hypothesis (12.48) holds for E. By Theorem 12.5(a) (Properties of Measurable Sets), $U \setminus E$ is measurable. Since U is a G_δ set, it is measurable. Since $E = U \setminus (U \setminus E)$, we conclude that E is also measurable and thus we have completed the proof of the problem.   ∎

Remark 12.6

Similar to the proof of Problem 12.33, we can show that E is measurable if and only if there exists a F_σ set W of $\mathbb{R}$ contained in E such that

$$m^*(E \setminus W) = 0 \tag{12.50}$$

Problem 12.34

⊛ ⊛ *Suppose that $E \subseteq \mathbb{R}$ is measurable with $m^*(E) < \infty$. For every $\epsilon > 0$, there is a finite disjoint collection of open intervals $\{I_1, I_2, \ldots, I_n\}$ such that if $I = \bigcup_{k=1}^{n} I_k$, then*

$$m^*(E \setminus I) + m^*(I \setminus E) < \epsilon.$$

Proof. By Problem 12.31, there exists an open set V in $\mathbb{R}$ such that

$$E \subseteq V \quad \text{and} \quad m^*(V \setminus E) < \frac{\epsilon}{2}. \tag{12.51}$$

Since $m^*(E) < \infty$, we get from this, the measurability of E and the inequality (12.51) that

$$m^*(V) = m^*(E) + m^*(V \setminus E) < \infty,$$

i.e., V has finite outer measure. Recall from [18, Exercise 29, p. 45] that there exists a countable disjoint collection of open intervals $\{I_k\}$ whose union is V. Since each I_k is measurable, it follows from Theorems 12.3(b) (Properties of Outer Measure) and 12.11 (Countably Additivity) that

$$\sum_{k=1}^{\infty} \ell(I_k) = \sum_{k=1}^{\infty} m(I_k) = m\Big(\bigcup_{k=1}^{\infty} I_k \Big) = m(V) = m^*(V) < \infty.$$

Now the definition of a series shows that there is a positive integer n such that

$$\sum_{k=n+1}^{\infty} \ell(I_k) < \frac{\epsilon}{2}. \tag{12.52}$$

Therefore, if we define $I = \bigcup_{k=1}^{n} I_k$, then since $I \setminus E \subseteq V \setminus E$, we know from Theorem 12.3(c) (Properties of Outer Measure) and the inequality (12.51) that

$$m^*(I \setminus E) \leq m^*(V \setminus E) < \frac{\epsilon}{2}. \tag{12.53}$$

Furthermore, since $E \subseteq V$, it is true that

$$E \setminus I \subseteq V \setminus I = \bigcup_{k=n+1}^{\infty} I_k.$$

Again we apply Theorems 12.3(b), (c), (d) (Properties of Outer Measure) and then using the inequality (12.52) to conclude that

$$m^*(E \setminus I) \leq m^* \Big(\bigcup_{k=n+1}^{\infty} I_k \Big) = \sum_{k=1}^{\infty} m^*(I_k) \leq \sum_{k=n+1}^{\infty} \ell(I_k) < \frac{\epsilon}{2}. \tag{12.54}$$

Hence our desired result follows directly from the sum of the two estimates (12.53) and (12.54). This completes the proof of the problem.[f]   ∎

Problem 12.35

$(\star)$ $(\star)$ The set A is said to be **Borel measurable** if A is a Borel set, i.e., $A \in \mathscr{B}$. Let $E \subseteq \mathbb{R}$. Prove that E is Lebesgue measurable if and only if

$$E = A \cup B,$$

where A is Borel measurable and $m^*(B) = 0$.

Proof. Suppose that $E = A \cup B$, where A is Borel measurable and $m^*(B) = 0$. By the paragraph following Theorem 12.10, A is mesurable. By Theorem 12.5(a) (Properties of Measurable Sets), B is measurable. Hence E is also measurable because it is an union of two measurable sets.

Conversely, we suppose that E is measurable. By Remark 12.6, there exists a F_σ set W in $\mathbb{R}$ such that $W \subseteq E$ and the formula (12.50) holds. Obviously, we have

$$E = W \cup (E \setminus W).$$

As a F_σ set, Theorem 12.10 guarantees that W belongs to $\mathscr{B}$ and thus W is Borel measurable. Besides, the formula (12.50) ensures that $E \setminus W$ acts as the role of the set B in the problem. Hence we complete the proof of the problem.   ∎

---

[f]This result is the first principle of the so-called **Littlewood's Three Principles**. For other two principles, please read Chapter 13.

CHAPTER 13

Lebesgue Measurable Functions

13.1 Fundamental Concepts

In Chapter 12, we review the basic definitions and results of Lebesgue measurable sets. Another building block of the theory is the main focus of this chapter: **Lebesgue measurable functions**. In this chapter, we are going to review the fundamental results of Lebesgue measurable functions, or for short measurable functions. The main references for this chapter are [5, Chap. 2], [9, Chap. 4, 5], [17, Chap. 3], [21, Chap. 1] and [22, Chap. 7].

13.1.1 Lebesgue Measurable Functions

Definition 13.1 (Lebesgue Measurable Functions)**.** *Denote* $\mathbb{R}^* = \mathbb{R} \cup \{\pm\infty\}$ *to be the extended real number system. The function* $f : \mathbb{R} \to \mathbb{R}^*$ *is called* **Lebesgue measurable**, *or simply* **measurable**, *if the set*

$$f^{-1}((a, \infty]) = \{x \in \mathbb{R} \mid f(x) > a\}$$

is measurable for all $a \in \mathbb{R}$.

Theorem 13.2 (Criteria for Measurable Functions)**.** *The following four conditions are equivalent.*

(a) *For every* $a \in \mathbb{R}$, *the set* $f^{-1}((a, \infty]) = \{x \in \mathbb{R} \mid f(x) > a\}$ *is measurable.*

(b) *For every* $a \in \mathbb{R}$, *the set* $f^{-1}([a, \infty]) = \{x \in \mathbb{R} \mid f(x) \geq a\}$ *is measurable.*

(c) *For every* $a \in \mathbb{R}$, *the set* $f^{-1}([-\infty, a)) = \{x \in \mathbb{R} \mid f(x) < a\}$ *is measurable.*

(d) *For every* $a \in \mathbb{R}$, *the set* $f^{-1}([-\infty, a]) = \{x \in \mathbb{R} \mid f(x) \leq a\}$ *is measurable.*

Remark 13.1

If the sets considered in Definition 13.1 (Lebesgue Measurable Functions) and Theorem 13.2 (Criteria for Measurable Functions) are Borel, then the function f is called **Borel measurable** or simply a **Borel function**.

Theorem 13.3. *The function* $f : \mathbb{R} \to \mathbb{R}$ *is Lebesgue (resp. Borel) measurable if and only if* $f^{-1}(V)$ *is Lebesgue (resp. Borel) measurable for every open set* V *in* $\mathbb{R}$. *Particularly, if* f *is continuous on* $\mathbb{R}$, *then* f *is Borel measurable.*

Similar to the situation of Lebesgue measurable sets and Borel measurable sets, a Borel measurable function must be Lebesgue measurable, but not the converse.

Theorem 13.4 (Properties of Measurable Functions). *Suppose that* $f, g : \mathbb{R} \to \mathbb{R}$ *are measurable,* $\{f_n\}$ *is a sequence of measurable functions and* $k \in \mathbb{N}$.

(a) *If* $\Phi : E \to \mathbb{R}$ *is continuous on* E, *where* $f(\mathbb{R}) \subseteq E$, *then the composition* $\Phi \circ f : \mathbb{R} \to \mathbb{R}$ *is also measurable.*[a]

(b) f^k, $f + g$ *and* fg *are measurable.*

(c) *The functions*

$$\sup_{n \in \mathbb{N}} f_n(x), \quad \inf_{n \in \mathbb{N}} f_n(x), \quad \limsup_{n \to \infty} f_n(x) \quad \text{and} \quad \liminf_{n \to \infty} f_n(x)$$

are all measurable.

(d) *If* $f(x) = h(x)$ *a.e. on* $\mathbb{R}$, *then* h *is measurable.*

13.1.2 Simple Functions and the Littlewood's Three Principles

Definition 13.5 (Simple Functions). *Let* $E \subseteq \mathbb{R}$. *The* **characteristic function** *of* E *is given by*

$$\chi_E(x) = \begin{cases} 1, & \text{if } x \in E; \\ 0, & \text{otherwise.} \end{cases}$$

If $E_1, E_2, \ldots, E_n$ *are pairwise disjoint measurable sets, then the function*

$$s(x) = \sum_{k=1}^{n} a_k \chi_{E_k}(x), \tag{13.1}$$

where $a_1, a_2, \ldots, a_n \in \mathbb{R}$, *is called a* **simple function**.

Theorem 13.6 (The Simple Function Approximation Theorem). *Suppose that* $f : \mathbb{R} \to \mathbb{R}$ *is measurable and* $f(x) \geq 0$ *on* $\mathbb{R}$. *Then there exists a sequence* $\{s_n\}$ *of simple functions such that*

$$0 \leq s_1(x) \leq s_2(x) \leq \cdots \leq f(x) \quad \text{and} \quad f(x) = \lim_{n \to \infty} s_n(x) \tag{13.2}$$

for all $x \in \mathbb{R}$.

> **Remark 13.2**
>
> The condition that f is nonnegative can be omitted in Theorem 13.6 (The Simple Function Approximation Theorem). In this case, the set of inequalities (13.2) are replaced by
>
> $$0 \leq |s_1(x)| \leq |s_2(x)| \leq \cdots \leq |f(x)|.$$

---

[a]Here the result also holds if Φ is a Borel function.

If the measurable sets $E_1, E_2, \ldots, E_n$ considered in the expression (13.1) are closed intervals, then the function s will be called a **step function**. In this case, we have the following approximation theorem in terms of step functions:

Theorem 13.7. *Suppose that* $f : \mathbb{R} \to \mathbb{R}$ *is measurable on* $\mathbb{R}$. *Then there exists a sequence* $\{\varphi_n\}$ *of step functions such that*

$$f(x) = \lim_{n \to \infty} \varphi_n(x)$$

a.e. on $\mathbb{R}$.

Classically, Littlewood says that (i) every measurable set is nearly a finite union of intervals; (ii) every measurable function is nearly continuous and (iii) every pointwise convergent sequence of measurable functions is nearly uniformly convergent. The mathematical formulation of the first principle has been given in Problem 12.34. Now the precise statements of the remaining principles are given in the following two results:

Theorem 13.8 (Egorov's Theorem). *Suppose that* $\{f_n\}$ *is a sequence of measurable functions defined on a measurable set* E *with finite measure. Given* $\epsilon > 0$. *If* $f_n \to f$ *pointwise a.e. on* E, *then there exists a closed set* $K_\epsilon \subseteq E$ *such that*

$$m(E \setminus K_\epsilon) < \epsilon \quad \text{and} \quad f_n \to f \text{ uniformly on } K_\epsilon.$$

Theorem 13.9 (Lusin's Theorem). *Suppose that* f *is a finite-valued measurable function on the measurable set* E *with* $m(E) < \infty$. *Given* $\epsilon > 0$. *Then there exists a closed set* $K_\epsilon \subseteq E$ *such that*

$$m(E \setminus K_\epsilon) < \epsilon \quad \text{and} \quad f \text{ is continuous on } K_\epsilon.$$

13.2 Lebesgue Measurable Functions

Problem 13.1

$(\star)$ *Suppose that* $f : \mathbb{R} \to \mathbb{R}$ *is Lebesgue measurable. Prove that* $|f|$ *is also Lebesgue measurable.*

Proof. Consider $\Phi : \mathbb{R} \to [0, \infty)$ defined by

$$\Phi(x) = |x|$$

which is clearly continuous. Since $|f| = \Phi \circ f$, Theorem 13.4(a) (Properties of Measurable Functions) ensures that $|f|$ is also Lebesgue measuable and it completes the proof of the problem. ∎

Problem 13.2

$(\star)$ *Suppose that* $f, g : \mathbb{R} \to \mathbb{R}$ *are measurable. Prove that* $\max(f, g)$ *and* $\min(f, g)$ *are measurable.*

Proof. Note that

$$\max(f, g) = \frac{1}{2}(f + g + |f - g|) \quad \text{and} \quad \min(f, g) = \frac{1}{2}(f + g - |f - g|).$$

By Problem 12.1 and Theorem 13.4(b) (Properties of Measurable Functions), we see immediately that both $\max(f, g)$ and $\min(f, g)$ are measurable. This completes the proof of the problem. ∎

Remark 13.3

As a particular case of Problem 13.2, the functions $f^+ = \max(f, 0)$ and $f^- = \min(f, 0)$ are measurable.

Problem 13.3

(⋆) Does there exist a nonmeasurable nonnegative function f such that $\sqrt{f}$ is measurable?

Proof. The answer is negative. Assume that $\sqrt{f}$ was measurable. We know that $\Phi(x) = x^2$ is continuous in $\mathbb{R}$. Now we have

$$f = (\sqrt{f})^2 = \Phi \circ \sqrt{f},$$

so Theorem 13.4(a) (Properties of Measurable Functions) ensures that f is measurable, a contradiction. Hence this completes the proof of the problem. ∎

Problem 13.4

(⋆) Suppose that $f : [0, \infty) \to \mathbb{R}$ is differentiable. Prove that f' is measurable.

Proof. Let $x \in [0, \infty)$. By the definition, we have

$$f(x) = \lim_{n \to \infty} f_n(x),$$

where $f_n(x) = \frac{f(x + \frac{1}{n}) - f(x)}{\frac{1}{n}}$. Since f is differentiable in $[0, \infty)$, it is continuous on $[0, \infty)$. By Theorem 13.3, f is (Borel) measurable.

For each $n \in \mathbb{N}$, we notice that

$$f\left(x + \frac{1}{n}\right) = (f \circ g)(x),$$

where $g(x) = x + \frac{1}{n}$ which is clearly continuous on $[0, \infty)$ and so measurable. Thus we follow from Theorem 13.4(a) (Properties of Measurable Functions)[b] that $f(x + \frac{1}{n})$ is measurable. By Theorem 13.4(b) and then (c) (Properties of Measurable Functions), each f_n and then f, as the limit of $\{f_n\}$, are also measurable. This ends the proof of the problem. ∎

---

[b] With Φ and f replaced by f and g in our question respectively.

Problem 13.5

(⋆) Suppose that $E \subseteq \mathbb{R}$ is measurable and $f : E \to \mathbb{R}$ is continuous on E. Prove that f is measurable.

Proof. Let $a \in \mathbb{R}$ and $A = \{x \in E \,|\, f(x) > a\}$. If $A = \varnothing$, then there is nothing to prove. Otherwise, pick $x \in A$. Then the Sign-preserving Property [25, Problem 7.15, p. 112] implies that there exists a $\delta_x > 0$ such that $f(y) > a$ holds for all $y \in (x - \delta_x, x + \delta_x) \cap E$. Therefore, we have

$$A = \bigcup_{x \in A} [(x - \delta_x, x + \delta_x) \cap E] = \Big[\bigcup_{x \in A} (x - \delta_x, x + \delta_x) \Big] \cap E.$$

It is trivial that the set

$$\bigcup_{x \in A} (x - \delta_x, x + \delta_x)$$

is open in $\mathbb{R}$, so it is measurable by Theorem 12.9. This fact and the measurability of E shows that A is measurable. By Definition 13.1 (Lebesgue Measurable Functions), f is measurable and this completes the proof of the problem. ∎

Problem 13.6

(⋆) If f is measurable on $E \subseteq \mathbb{R}$, prove that $\{x \in E \,|\, f(x) = a\}$ is measurable for every $a \in \mathbb{R}$.

Proof. We have

$$\{x \in E \,|\, f(x) = a\} = \{x \in E \,|\, f(x) \geq a\} \cap \{x \in E \,|\, f(x) \leq a\}. \tag{13.3}$$

By Theorem 13.4 (Properties of Measurable Functions), the sets on the right-hand side of the expression (13.3) are measurable. Finally, we apply Theorem 12.5(e) (Properties of Measurable Sets) to conclude that

$$\{x \in E \,|\, f(x) = a\}$$

is also measurable, so it ends the analysis of the problem. ∎

Problem 13.7

(⋆) If f is measurable on $E \subseteq \mathbb{R}$, prove that $\{x \in E \,|\, f(x) = \infty\}$ is measurable.

Proof. We observe that

$$\{x \in E \,|\, f(x) = \infty\} = \bigcap_{n=1}^{\infty} \{x \in E \,|\, f(x) > n\}. \tag{13.4}$$

Since f is measurable on E, each set $\{x \in E \,|\, f(x) > n\}$ is measurable. Recall from Remark 12.3 that the set on the right-hand side of the expression (13.4) is measurable and our desired result follows. This completes the proof of the problem. ∎

Problem 13.8

⭐ *Suppose that f and g are measurable on $E \subseteq \mathbb{R}$. Prove that the set $\{x \in E \mid f(x) > g(x)\}$ is measurable.*

Proof. For every $x \in E$, since $f(x) > g(x)$, the density of rationals [25, Theorem 2.2, p. 10] implies that there is a $r \in \mathbb{Q}$ such that $f(x) > r > g(x)$. Therefore, we have

$$\{x \in E \mid f(x) > r > g(x)\} = \{x \in E \mid f(x) > r\} \cap \{x \in E \mid g(x) < r\}. \tag{13.5}$$

By the hypotheses, both sets $\{x \in E \mid f(x) > r\}$ and $\{x \in E \mid g(x) < r\}$ are measurable. Thus it follows from this and the expression (13.5) that the set

$$\{x \in E \mid f(x) > r > g(x)\}$$

is also measurable. Now we know that

$$\{x \in E \mid f(x) > g(x)\} = \bigcup_{r \in \mathbb{Q}} \{x \in E \mid f(x) > r > g(x)\}$$

and then Theorem 12.8 guarantees that $\{x \in E \mid f(x) > g(x)\}$ is measurable and we complete the analysis of the problem. ∎

Problem 13.9

⭐ *Prove that*

(a) *a characteristic function χ_E is measurable if and only if E is measurable.*

(b) *a simple function $s = \sum_{k=1}^{n} a_k \chi_{E_k}$ is measurable.*

Proof. In the following discussion, suppose that $E, E_1, \ldots, E_n$ are measurable.

(a) We deduce from Definition 13.5 (Simple Functions) that

$$\{x \in \mathbb{R} \mid \chi_E(x) < a\} = \begin{cases} \varnothing, & \text{if } a \le 0; \\ \mathbb{R}, & \text{if } a > 1; \\ \mathbb{R} \setminus E, & \text{if } 0 < a \le 1. \end{cases}$$

Consequently, the set $\{x \in \mathbb{R} \mid \chi_E(x) < a\}$ is measurable for every $a \in \mathbb{R}$ and Definition 13.1 (Lebesgue Measurable Functions) implies that χ_E is measurable. Conversely, suppose that χ_E is measurable. Since $E = \mathbb{R} \setminus \{x \in \mathbb{R} \mid \chi_E(x) < \frac{1}{2}\}$ and $\{x \in \mathbb{R} \mid \chi_E(x) < \frac{1}{2}\}$ is measurable, we conclude that E is also measurable.

(b) By part (a), each χ_{E_k} is measurable. Then repeated applications of Theorem 13.4(b) (Properties of Measurable Functions) show that the simple function s is also measurable.

We complete the analysis of the proof.   ∎

> **Problem 13.10**
>
> (⋆) *Construct a nonmeasurable function.*

Proof. By Remark 12.2, $\mathbb{R}$ contains a nonmeasurable set S. Consider the characteristic function

$$\chi_S(x) = \begin{cases} 1, & \text{if } x \in S; \\ 0, & \text{otherwise.} \end{cases}$$

Then χ_S cannot be measurable by Problem 13.9. This completes the proof of the problem.   ∎

> **Problem 13.11**
>
> (⋆) *Suppose that $f : \mathbb{R} \to (0, \infty)$ is measurable. Prove that $\frac{1}{f}$ is measurable.*

Proof. Let $g = \frac{1}{f} : \mathbb{R} \to (0, \infty)$. If $a \le 0$, then $\{x \in \mathbb{R} \mid g(x) > a\} = \mathbb{R}$ which is definitely measurable. For $a > 0$, we note that $g(x) > a$ if and only if $0 < f(x) < \frac{1}{a}$ so that

$$\begin{aligned} \{x \in \mathbb{R} \mid g(x) > a\} &= \left\{ x \in \mathbb{R} \,\middle|\, 0 < f(x) < \frac{1}{a} \right\} \\ &= \{x \in \mathbb{R} \mid f(x) > 0\} \cap \left\{ x \in \mathbb{R} \,\middle|\, f(x) < \frac{1}{a} \right\}. \end{aligned} \tag{13.6}$$

Since f is measurable, Theorem 13.2 (Criteria for Measurable Functions) ensures that the two sets on the right-hand side of the expression (13.6) are measurable. Thus the set

$$\{x \in \mathbb{R} \mid g(x) > a\}$$

is measurable. By Definition 13.1 (Lebesgue Measurable Functions), the g is a measurable function, completing the proof of the problem.   ∎

> **Problem 13.12**
>
> (⋆) (⋆) *Suppose that $f : \mathbb{R} \to \mathbb{R}$ is measurable and $f(x + 1) = f(x)$ a.e. on $\mathbb{R}$. Prove that there exists a function $g : \mathbb{R} \to \mathbb{R}$ such that $f = g$ a.e. and $g(x + 1) = g(x)$ on $\mathbb{R}$.*

Proof. Suppose that $E = \{x \in \mathbb{R} \mid f(x + 1) \neq f(x)\}$. By the hypothesis, we have $m(E) = 0$. Next, we define

$$F = \bigcup_{n \in \mathbb{Z}} (E + n) \tag{13.7}$$

and

$$g(x) = \begin{cases} f(x), & \text{if } x \notin F; \\ 0, & \text{otherwise.} \end{cases} \tag{13.8}$$

By Theorem 12.5(d) (Properties of Measurable Sets), each $E+n$ is measurable and $m(E+n) = 0$. Applying this fact to the inequality (12.1), we derive

$$m(F) \leq \sum_{n=1}^{\infty} m(E+n) = 0.$$

Therefore, we follow from the definition (13.8) that

$$f = g$$

a.e. on $\mathbb{R}$, proving the first assertion.

Next, since $g(x+1) = g(x)$ on F^c, it suffices to prove the second assertion on F. To see this, if $x \in F$, then we note from the definition (13.7) that $x = E + m$ for some $m \in \mathbb{Z}$ which implies that

$$x + 1 = E + m + 1 \in F.$$

Hence it is obvious from the definition (13.8) that for every $x \in F$, we still have

$$g(x + 1) = 0 = g(x)$$

which proves the second assertion and this completes the proof of the problem. ∎

Problem 13.13

⋆ ⋆ *Let $f : \mathbb{R} \to \mathbb{R}$ be continuous a.e. on $\mathbb{R}$. Prove that f is Lebesgue measurable.*

Proof. Let $a \in \mathbb{R}$ and denote $E_a = \{x \in \mathbb{R} \mid f(x) > a\}$. Let $x \in E_a$ and A be the set of all discontinuities of f. There are two cases:

- **Case (1): f is discontinuous at x.** In this case, $A \neq \varnothing$. By the hypothesis, we know that $m(A) = 0$ so that A is measurable by Theorem 12.5(a) (Properties of Measurable Sets).

- **Case (2): f is continuous at x.** In this case, the Sign-preserving Property [25, Problem 7.15, p. 112] ensures that there is a $\delta_x > 0$ such that $f(y) > a$ for all $y \in (x - \delta_x, x + \delta_x)$. In other words, we have $(x - \delta_x, x + \delta_x) \subseteq E_a$.

Therefore, we are able to write

$$E_a = A \cup \bigcup_{x \in E_a \setminus A} (x - \delta_x, x + \delta_x).$$

Obviously, since every $(x - \delta_x, x + \delta_x)$ is open in $\mathbb{R}$, the union

$$\bigcup_{x \in E_a \setminus A} (x - \delta_x, x + \delta_x) \tag{13.9}$$

is also open in $\mathbb{R}$. By Theorem 12.9, the set (13.9) is measurable. Hence we deduce from **Case (1)** that E_a is also measurable. Recall that a is arbitrary, so Theorem 13.2 (Criteria for Measurable Functions) concludes that f is Lebesgue measurable which completes the proof of the problem. ∎

Problem 13.14

$\bigstar$ $\bigstar$ Let $f : \mathbb{R} \to \mathbb{R}$ be monotone. Verify that f is Borel measurable.

Proof. Without loss of generality, we may assume that f is increasing. Let $a \in \mathbb{R}$. Suppose that $p \in \{x \in \mathbb{R} \mid f(x) < a\}$. Then $f(p) < a$ and for all $y < p$, since f is increasing, we have

$$f(y) < f(p) < a.$$

Thus we have $y \in \{x \in \mathbb{R} \mid f(x) < a\}$. Now we put $\alpha = \sup\{x \in \mathbb{R} \mid f(x) < a\}$. On the one hand, if $f(\alpha) < a$, then we have

$$\{x \in \mathbb{R} \mid f(x) < a\} = (-\infty, \alpha].$$

On the other hand, if $f(\alpha) \geq a$, then we have

$$\{x \in \mathbb{R} \mid f(x) < a\} = (-\infty, \alpha).$$

In any case, we conclude from Theorem 12.10(c) that the set $\{x \in \mathbb{R} \mid f(x) < a\}$ is Borel measurable. By Remark 13.1, the function f is a Borel function, completing the proof of the problem. ∎

Problem 13.15

$\bigstar$ $\bigstar$ Let S be dense in $\mathbb{R}$ and $f : \mathbb{R} \to \mathbb{R}$. Prove that f is measurable if and only if the set $\{x \in \mathbb{R} \mid f(x) > s\}$ is measurable for every $s \in S$.

Proof. Suppose first that f is measurable. Then Theorem 13.2 (Criteria for Measurable Functions) says that $\{x \in \mathbb{R} \mid f(x) > a\}$ is Lebesgue measurable for every $a \in \mathbb{R}$. In particular, the set $\{x \in \mathbb{R} \mid f(x) > s\}$ is measurable for every $s \in S$.

For the converse direction, it suffices to prove that $\{x \in \mathbb{R} \mid f(x) > s\}$ is measurable for every $s \in S^c$. In fact, given $p \in S^c$. Since S is dense in $\mathbb{R}$, there exists a decreasing sequence $\{p_n\} \subseteq S$ such that $p_n \to p$ as $n \to \infty$. Next, we claim that

$$\{x \in \mathbb{R} \mid f(x) > p\} = \bigcup_{n=1}^{\infty} \{x \in \mathbb{R} \mid f(x) > p_n\}. \tag{13.10}$$

To see this, for each $n \in \mathbb{N}$, if $y \in \{x \in \mathbb{R} \mid f(x) > p_n\}$, then since $p_n > p$, we have $f(y) > p$ so that $y \in \{x \in \mathbb{R} \mid f(x) > p\}$, i.e.,

$$\bigcup_{n=1}^{\infty} \{x \in \mathbb{R} \mid f(x) > p_n\} \subseteq \{x \in \mathbb{R} \mid f(x) > p\}. \tag{13.11}$$

On the other hand, if $y \in \{x \in \mathbb{R} \mid f(x) > p\}$, then we have $\epsilon = f(y) - p > 0$. Since p_n converges to p decreasingly, there exists an $N \in \mathbb{N}$ such that $n \geq N$ implies $p_n - p < \epsilon$ and this means that

$$f(y) > p_n,$$

i.e., $y \in \{x \in \mathbb{R} \mid f(x) > p_n\}$ for all $n \geq N$. Consequently, we have

$$\{x \in \mathbb{R} \mid f(x) > p\} \subseteq \bigcup_{n=1}^{\infty} \{x \in \mathbb{R} \mid f(x) > p_n\}. \tag{13.12}$$

Now our claim (13.10) follows immediately by combining the set relations (13.11) and (13.12). By the assumption, each $\{x \in \mathbb{R} \mid f(x) > p_n\}$ is measurable, so the expression (13.10) shows that

$$\{x \in \mathbb{R} \mid f(x) > p\}$$

is also measurable, where $p \in S^c$. By Definition 13.1 (Lebesgue Measurable Functions), f is measurable and it completes the analysis of the problem. ∎

Problem 13.16

$(\star)(\star)$ *Suppose that $f : \mathbb{R} \to \mathbb{R}$ is bijective and continuous and $\mathscr{B}$ denotes the collection of all Borel sets of $\mathbb{R}$. Prove that $f(\mathscr{B}) \subseteq \mathscr{B}$.*

Proof. Suppose that

$$\mathscr{D} = \{E \subseteq \mathbb{R} \mid f(E) \in \mathscr{B}\}.$$

First of all, we want to show that $\mathscr{D}$ is an σ-algebra. To this end, recall a basic fact from [12, Exercise 2(g) and (h), p. 21] that if f is bijective, then

$$f(E \cap F) = f(E) \cap f(F) \quad \text{and} \quad f(E \setminus F) = f(E) \setminus f(F). \tag{13.13}$$

Thus if $E \in \mathscr{D}$, then the second set equation (13.13) implies that

$$f(E^c) = f(\mathbb{R}) \setminus f(E) = \mathbb{R} \setminus f(E).$$

Since $f(E), \mathbb{R} \in \mathscr{B}$, we have $f(E^c) \in \mathscr{B}$ and then $E^c \in \mathscr{D}$. Let $E_k \in \mathscr{D}$ so that

$$f\left(\bigcup_{k=1}^{\infty} E_k\right) = \bigcup_{k=1}^{\infty} f(E_k). \tag{13.14}$$

Since $f(E_k) \in \mathscr{B}$ for every $k = 1, 2, \ldots$, we have

$$\bigcup_{k=1}^{\infty} f(E_k) \in \mathscr{B}$$

and then the set relation (13.14) shows that

$$\bigcup_{k=1}^{\infty} E_k \in \mathscr{D}.$$

Therefore, $\mathscr{D}$ is an σ-algebra by Definition 12.7 (σ-algebra) which proves our claim.

Next, for every closed interval $[a, b]$, the continuity of f ensures that $f([a, b])$ is compact and connected by [25, Theorems 7.9 and 7.12, p. 100]. Note that $f([a, b]) \subseteq \mathbb{R}$, so [25, Theorem 4.16,

p. 31] implies that $f([a,b])$ must be a closed interval. Since f is continuous and one-to-one, we know from [25, Problem 7.33, p. 122] that f is strictly monotonic and hence either

$$f([a,b]) = [f(a), f(b)] \quad \text{or} \quad f([a,b]) = [f(b), f(a)].$$

Now both $[f(a), f(b)]$ and $[f(b), f(a)]$ are closed in $\mathbb{R}$, so they belong to $\mathscr{B}$. By the definition, we conclude that $[a,b] \in \mathscr{D}$ for every $-\infty < a \le b < \infty$. In other words, $\mathscr{D}$ contains every closed intervals of $\mathbb{R}$ and Theorem 12.10(c) says that $\mathscr{B} \subseteq \mathscr{D}$. Since $\mathscr{D}$ is an σ-algebra and $\mathscr{B}$ is a **smallest** σ-algebra containing all closed intervals of $\mathbb{R}$, we establish that

$$\mathscr{B} = \mathscr{D}$$

which means $f(\mathscr{B}) \subseteq \mathscr{B}$, as desired. Hence we complete the analysis of the problem. ∎

Problem 13.17

⭐ *Let E be measurable. Prove that $f : E \to \mathbb{R}$ is measurable if and only if $f^{-1}(V)$ is measurable for every open set V in $\mathbb{R}$.*

Proof. We suppose that $f^{-1}(V)$ is measurable for every open set V in $\mathbb{R}$. In particular,

$$f^{-1}((a, \infty)) = \{x \in E \mid f(x) > a\}$$

is measurable for every $a \in \mathbb{R}$. By Theorem 13.2 (Criteria for Measurable Functions), f is measurable.

Conversely, suppose that f is measurable. By [18, Exercise 29, p. 45], V is an union of countable collection of disjoint open intervals $\{(a_n, b_n)\}$, i.e.,

$$V = \bigcup_{n=1}^{\infty} (a_n, b_n).$$

Recall the two facts[c]

$$f^{-1}(A \cup B) = f^{-1}(A) \cup f^{-1}(B) \quad \text{and} \quad f^{-1}(A \cap B) = f^{-1}(A) \cap f^{-1}(B),$$

we have

$$
\begin{aligned}
f^{-1}(V) &= \bigcup_{n=1}^{\infty} f^{-1}((a_n, b_n)) \\
&= \bigcup_{n=1}^{\infty} f^{-1}((-\infty, b_n) \cap (a_n, \infty)) \\
&= \bigcup_{n=1}^{\infty} f^{-1}((-\infty, b_n)) \cap f^{-1}((a_n, \infty)).
\end{aligned}
$$

By Theorem 13.2 (Criteria for Measurable Functions), both $f^{-1}((-\infty, b_n))$ and $f^{-1}((a_n, \infty))$ are measurable so that every $f^{-1}((a_n, b_n))$ is measurable. Thus Theorem 12.8 implies that $f^{-1}(V)$ is measurable, as desired. Hence we have completed the proof of the problem. ∎

---

[c]See [12, Exercise 2(b) and (c), p. 20].

Problem 13.18

$(\star)(\star)$ *Suppose that $E \subseteq \mathbb{R}$ is measurable and $m(E) < \infty$. Let f be measurable on E and $|f(x)| < \infty$ a.e. on E. Prove that for every $\epsilon > 0$, there exists a measurable subset $F \subseteq E$ such that*

$$m(E \setminus F) < \epsilon \quad \text{and} \quad f|_F \text{ is bounded on } F.$$

Proof. For each $n \in \mathbb{N}$, consider

$$E_n = \{x \in E \,|\, |f(x)| > n\} \tag{13.15}$$

and $A = \{x \in E \,|\, f(x) = \pm\infty\}$. Since $|f(x)| < \infty$ a.e. on E, we have $m(A) = 0$. Since f is measurable on E and

$$E_n = \{x \in E \,|\, f(x) > n\} \cup \{x \in E \,|\, f(x) < -n\},$$

every E_n is also measurable. Furthermore, it is easy to check from the definition (13.15) that

$$E_1 \supseteq E_2 \supseteq \cdots \quad \text{and} \quad A = \bigcap_{n=1}^{\infty} E_n.$$

Since $m(E_1) \leq m(E) < \infty$, we apply Theorem 12.12 (Continuity of Lebesgue Measure) to conclude that

$$\lim_{n \to \infty} m(E_n) = m(A) = 0. \tag{13.16}$$

Thus given $\epsilon > 0$, there exists an $N \in \mathbb{N}$ such that $m(E_N) < \epsilon$. Put $F = E \setminus E_N$. Then we have

$$E_N = E \setminus F \quad \text{and} \quad F = \{x \in E \,|\, |f(x)| \leq N\}$$

which mean that $f|_F$ is definitely bounded by N on F. Therefore, this completes the proof of the problem. ∎

Problem 13.19

$(\star)(\star)$ *Suppose that $\{f_n\}$ is a sequence of measurable functions on $[0,1]$ with $|f_n(x)| < \infty$ a.e. on $[0,1]$. Prove that there exists a sequence $\{a_n\}$ of positive real numbers such that*

$$\frac{f_n(x)}{a_n} \to 0 \tag{13.17}$$

as $n \to \infty$ a.e. on $[0,1]$.

Proof. Given $\epsilon > 0$. Using the same notations and argument as in Problem 13.18, we still have the limit (13.16). In other words, for each $n \in \mathbb{N}$, there corresponds a $k_n \in \mathbb{N}$ such that

$$m(E_k(n)) = m(\{x \in [0,1] \,|\, |f_n(x)| > k_n\}) < \epsilon, \tag{13.18}$$

where

$$E_k(n) = \{x \in [0,1] \,|\, |f_n(x)| > k\}.$$

Now we pick $\epsilon = 2^{-n}$ and $a_n = nk_n > 0$ in the estimate (13.18) to get

$$m(E_k(n)) = m\left(\left\{x \in [0,1] \,\Big|\, \frac{|f_n(x)|}{a_n} > \frac{1}{n}\right\}\right) < 2^{-n}$$

which implies that

$$\sum_{n=1}^{\infty} m(E_k(n)) < \sum_{n=1}^{\infty} 2^{-n} = 1 < \infty.$$

Hence it follows from Problem 12.15 (The Borel-Cantelli Lemma) that the set

$$E = \{x \in [0,1] \,|\, x \in E_k(n) \text{ for infinitely many } n\}$$

has measure 0. In other words, every $x \in E^c$ lies in *finitely many* $E_k(n)$. Thus, if $x \in E^c$, then there exists an $N \in \mathbb{N}$ such that $n \geq N$ implies

$$\frac{|f_n(x)|}{a_n} \leq \frac{1}{n}$$

and this gives the limit (13.17) a.e. on $[0,1]$ because $m(E^c) = 1$. We end the proof of the problem. ∎

Problem 13.20

(⋆)(⋆) *Suppose that $f : \mathbb{R} \to \mathbb{R}$ is measurable and V is an open set in $\mathbb{R}$ containing 0. Prove that there exists a measurable set E such that $m(E) > 0$ and $f(x) - f(y) \in V$ for every pair $x, y \in E$.*

Proof. Since V is an open set in $\mathbb{R}$ containing 0, there exists a $\epsilon > 0$ such that $(-\epsilon, \epsilon) \subseteq V$. Let $I = (-\frac{\epsilon}{2}, \frac{\epsilon}{2})$ and

$$E_p = f^{-1}(p + I) \tag{13.19}$$

for every $p \in \mathbb{R}$. Note that every E_p is measurable by Theorem 13.3.

We claim that $\bigcup_{p \in \mathbb{Q}} (p + I) = \mathbb{R}$. Otherwise, there was a $x \in \mathbb{R}$ such that

$$x \notin \bigcup_{p \in \mathbb{Q}} (p + I).$$

However, there exists a rational q such that $q \in (x, x + \frac{\epsilon}{4})$ and this implies that

$$x \in \left(q - \frac{\epsilon}{2}, q + \frac{\epsilon}{2}\right).$$

Therefore, no such x exists and we have the claim. By this, we obtain

$$\bigcup_{p \in \mathbb{Q}} E_p = \bigcup_{p \in \mathbb{Q}} f^{-1}(p + I) = f^{-1}\left(\bigcup_{p \in \mathbb{Q}} (p + I)\right) = f^{-1}(\mathbb{R}) = \mathbb{R}. \tag{13.20}$$

Assume that $m(E_p) = 0$ for all $p \in \mathbb{Q}$. Then it follows from the set relation (13.20) and the inequality (12.1) that

$$\infty = m(\mathbb{R}) \leq \sum_{p \in \mathbb{Q}} m(E_p) = 0,$$

a contradiction. Thus there exists a $p_0 \in \mathbb{Q}$ such that $m(E_{p_0}) > 0$ which implies that $E_{p_0} \neq \varnothing$. Now for every pair $x, y \in E_{p_0}$, the definition (13.19) shows that $f(x), f(y) \in (p_0 - \frac{\epsilon}{2}, p_0 + \frac{\epsilon}{2})$ which certainly gives

$$f(x) - f(y) \in (-\epsilon, \epsilon).$$

It completes the proof of the problem. ∎

13.3 Applications of Littlewood's Three Principles

Problem 13.21

(⋆)(⋆)(⋆) *Prove Theorem 13.8 (Egorov's Theorem).*

Proof. For each pair of $s, t \in \mathbb{N}$, we consider

$$E_t(s) = \left\{ x \in E \,\middle|\, |f_i(x) - f(x)| < \tfrac{1}{s} \text{ for all } i \geq t \right\}.$$

We fix s. Notice that

$$E_1(s) \subseteq E_2(s) \subseteq \cdots \quad \text{and} \quad \lim_{t \to \infty} E_t(s) = E.$$

By Theorem 12.12 (Continuity of Lebesgue Measure), we see that there exists a $t_s \in \mathbb{N}$ such that

$$m(E) - m(E_{t_s}(s)) < \frac{1}{2^s}. \tag{13.21}$$

Using Problem 12.13, we can express the inequality (13.21) as

$$m(E \setminus E_{t_s}(s)) < \frac{1}{2^s}.$$

Now we can choose $N \in \mathbb{N}$ large enough so that

$$\sum_{s=N}^{\infty} \frac{1}{2^s} < \frac{\epsilon}{2}. \tag{13.22}$$

Suppose that

$$K_{\epsilon,N} = \bigcap_{s=N}^{\infty} E_{t_s}(s)$$

which is measurable.

Now we are ready to prove the requirements of the theorem. In fact, we find from the inequalities (12.1) and (13.22) that

$$m(E \setminus K_{\epsilon,N}) = m\left(\bigcup_{s=N}^{\infty} (E \setminus E_{t_s}(s)) \right) \leq \sum_{s=N}^{\infty} m(E \setminus E_{t_s}(s)) < \frac{\epsilon}{2}. \tag{13.23}$$

Since $K_{\epsilon,N}$ is measurable, Remark 12.5 ensures that there is a closed set $K_\epsilon \subseteq K_{\epsilon,N}$ such that

$$m(K_{\epsilon,N} \setminus K_\epsilon) < \frac{\epsilon}{2}. \tag{13.24}$$

By combining the inequalities (13.23) and (13.24), we establish that

$$m(E \setminus K_\epsilon) = m((E \setminus K_{\epsilon,N}) \cup (K_{\epsilon,N} \setminus K)) \leq m(E \setminus K_{\epsilon,N}) + m(K_{\epsilon,N} \setminus K) < \epsilon.$$

This gives the first requirement of the theorem.

Next, we observe that $x \in K_{\epsilon,N}$ implies that $x \in E_{t_s}(s)$ for all $s \geq N$. Thus given any $\delta > 0$, we can find an $N' \geq N$ such that $\frac{1}{N'} < \delta$. Therefore, if $i \geq t_{N'}$, then the definition implies

$$|f_i(x) - f(x)| < \frac{1}{N'} < \delta \tag{13.25}$$

on $K_{\epsilon,N}$. Recall that $K_\epsilon \subseteq K_{\epsilon,N}$, so our result (13.25) is also valid on K_ϵ. This proves the second requirement of the theorem and it completes the proof of the problem. ∎

Problem 13.22

⋆ *Can we drop the condition $m(E) < \infty$ in Theorem 13.8 (Egorov's Theorem)?*

Proof. The answer is negative. In fact, consider $E = [0, \infty)$ and $f_n = \chi_{[n,\infty)} : E \to \mathbb{R}$. Since $[n, \infty)$ is measurable, f_n is measurable by Problem 13.9. For every $x \in E$, we have

$$f_n(x) \to f(x) = 0$$

as $n \to \infty$. Assume that there was a closed set $K \subseteq E$ such that

$$m(E \setminus K) < 1 \quad \text{and} \quad f_n \to 0 \text{ uniformly on } K. \tag{13.26}$$

Since $m(E) = \infty$, it is true that $m(K) = \infty$ so that K is unbounded. Thus for every $n \in \mathbb{N}$, we have

$$K \cap [n, \infty) \neq \varnothing.$$

Let $p_n \in K \cap [n, \infty)$. Then we have

$$f_n(p_n) = \chi_{[n,\infty)}(p_n) = 1$$

for every $n \in \mathbb{N}$ which contradicts the second condition (13.26). This completes the proof of the problem. ∎

Problem 13.23

⋆ *Suppose that $\{f_n\}$ is a sequence of measurable functions defined on a measurable set E with $m(E) < \infty$ and $f_n \to f$ pointwise a.e. on E. Prove that there exists a sequence $\{F_k\}$ of closed sets of E such that*

$$m\left(E \setminus \bigcup_{k=1}^{\infty} F_k\right) = 0 \quad \text{and} \quad f_n \to f \text{ uniformly on } F_1, F_2, \ldots.$$

Proof. If $m(E) < \infty$, then Theorem 13.8 (Egorov's Theorem) implies that there is a closed set $F_1 \subseteq E$ such that

$$m(E \setminus F_1) < \frac{1}{2} \quad \text{and} \quad f_n \to f \text{ uniformly on } F_1.$$

This argument can be applied repeatedly. In fact, for every $k \in \mathbb{N}$, there is a sequence $\{F_1, F_2, \ldots, F_k\}$ of closed sets with $F_k \subseteq E \setminus \bigcup_{i=1}^{k-1} F_i$ such that

$$m\left(E \setminus \bigcup_{i=1}^{k} F_i\right) < \frac{1}{2^k} \quad \text{and} \quad f_n \to f \text{ uniformly on } F_1, F_2, \ldots, F_k. \tag{13.27}$$

Define $F = \bigcup_{i=1}^{\infty} F_i$. Then it is easy to check that

$$m(E \setminus F) = m\left(E \setminus \bigcup_{i=1}^{\infty} F_i\right) \leq m\left(E \setminus \bigcup_{i=1}^{k} F_i\right)$$

for every $k \in \mathbb{N}$, so we further deduce from the inequality (13.27) that

$$m(E \setminus F) < \frac{1}{2^k} \tag{13.28}$$

for every $k \in \mathbb{N}$. Taking $k \to \infty$ in the inequality (13.28), we may conclude that

$$m(E \setminus F) = 0$$

which completes the proof of the problem. ∎

Problem 13.24

$(\star)$ Suppose that $\{f_n\}$ is a sequence of measurable functions defined on a measurable set E with finite measure. We say $\{f_n\}$ **converges in measure** to the measurable function f if for every $\epsilon > 0$, there corresponds an $N \in \mathbb{N}$ such that

$$m(\{x \in E \mid |f_n(x) - f(x)| > \epsilon\}) < \epsilon$$

for all $n > N$. Prove that if $f_n \to f$ pointwise a.e. on E, then $f_n \to f$ in measure.

Proof. Given that $\epsilon > 0$. Let $A = \{x \in E \mid f_n(x) \to f(x)\}$. Then we have $m(E \setminus A) = 0$. Since $m(E) < \infty$, we have $m(A) < \infty$ and we are able to apply Theorem 13.8 (Egorov's Theorem). Therefore, there is a closed set $K_\epsilon \subseteq A$ such that

$$m(A \setminus K_\epsilon) < \epsilon \quad \text{and} \quad f_n \to f \text{ uniformly on } K_\epsilon.$$

By the definition of uniform convergence, there exists an $N \in \mathbb{N}$ such that $n > N$ implies that

$$|f_n(x) - f(x)| < \epsilon$$

on K_ϵ. In other words, if $n > N$, then $|f_n(x) - f(x)| > \epsilon$ *only possibly* on $A \setminus K_\epsilon$ or on $E \setminus A$. Hence, for all $n > N$, we obtain

$$m(\{x \in E \mid |f_n(x) - f(x)| > \epsilon\}) \leq m(A \setminus K_\epsilon) + m(E \setminus A) < \epsilon.$$

By the definition, $f_n \to f$ in measure and it completes the proof of the problem. ∎

Problem 13.25

$(\star)(\star)$ *Prove Theorem 13.9 (Lusin's Theorem) in the special case that f is a simple function defined on E.*

Proof. Suppose that f takes the form

$$f(x) = \sum_{i=1}^{n} a_i \chi_{E_i}(x), \tag{13.29}$$

where $a_1, a_2, \ldots, a_n$ are distinct and $E_1, E_2, \ldots, E_n$ are pairwise disjoint subsets of E whose union is E. By Remark 12.5, there exist closed sets $K_1, K_2, \ldots, K_n$ such that

$$K_i \subseteq E_i \quad \text{and} \quad m(E_i \setminus K_i) < \frac{\epsilon}{n},$$

where $i = 1, 2, \ldots, n$. Set $K_\epsilon = \bigcup_{i=1}^{n} K_i$ which is clearly closed in $\mathbb{R}$. Then we deduce from the inequality (12.1) that

$$m(E \setminus K_\epsilon) = m\Big(\bigcup_{i=1}^{n} (E_i \setminus K_i) \Big) \leq \sum_{i=1}^{n} m(E_i \setminus K_i) < \epsilon.$$

This gives our first assertion.

For the second assertion, we define $g : K_\epsilon \to \mathbb{R}$ by $g(x) = a_i$ for $x \in K_i$. In fact, g takes the form

$$g(x) = \sum_{i=1}^{n} a_i \chi_{K_i}(x). \tag{13.30}$$

By comparing the two representations (13.29) and (13.30), we find that

$$f_{K_\epsilon} = g,$$

so it suffices to show that g is continuous on K_ϵ. To this end, consider $p \in K_i$ and denote $\widehat{K_i} = \bigcup_{\substack{j=1 \\ j \neq i}}^{n} K_j$. We claim that there exists a $\delta > 0$ such that

$$(p - \delta, p + \delta) \cap \widehat{K_i} = \varnothing.$$

Otherwise, for each $s \in \mathbb{N}$, we have

$$x_s \in \Big(p - \frac{1}{s}, p + \frac{1}{s} \Big) \cap \widehat{K_i}$$

so that $x_s \to p$ as $s \to \infty$. Since the set $\widehat{K_i}$ is closed in $\mathbb{R}$, it is also true that $p \in \widehat{K_i}$. This contradicts the fact that $K_1, K_2, \ldots, K_n$ are pairwise disjoint. Therefore, we have

$$(p - \delta, p + \delta) \cap K_\epsilon = (p - \delta, p + \delta) \cap K_i$$

so that

$$g(x) = a_i \tag{13.31}$$

for all $x \in (p-\delta, p+\delta) \cap K_\epsilon$. Since $(p-\delta, p+\delta) \cap K_\epsilon$ is equivalent to $|x-p| < \delta$ and $x \in K_\epsilon$. Hence the result (13.31) actually means that g is continuous at p. As K_i and then p are arbitrary, we have shown that g is continuous on K_ϵ and thus our second assertion is proven. Hence we have completed the proof of the problem.   ∎

Problem 13.26

 Prove Theorem 13.9 (Lusin's Theorem) with the aid of Problem 13.25.

Proof. Given $\epsilon > 0$. By Theorem 13.6 (The Simple Function Approximation Theorem), we see that there corresponds a sequence $\{s_n\}$ of simple functions defined on E converging to f pointwise on E. For each $n \in \mathbb{N}$, Problem 13.25 ensures that there exists a closed set $K_n \subseteq E$ such that

$$m(E \setminus K_n) < \frac{\epsilon}{2^{n+1}} \quad \text{and} \quad s_n \text{ is continuous on } K_n. \tag{13.32}$$

Since $s_n \to f$ pointwise on E, it follows from Theorem 13.8 (Egorov's Theorem) that there is a closed set $K_0 \subseteq E$ such that

$$m(E \setminus K_0) < \frac{\epsilon}{2} \quad \text{and} \quad s_n \to f \text{ uniformly on } K_0. \tag{13.33}$$

Now we set

$$K_\epsilon = \bigcap_{n=0}^{\infty} K_n.$$

By De Morgan's Laws and then the applications of the inequality (12.1), (13.32) and (13.33), we obtain

$$m(E \setminus K_\epsilon) = m\left((E \setminus K_0) \cup \bigcup_{n=1}^{\infty} (E \setminus K_n)\right) \leq m(E \setminus K_0) + \sum_{n=1}^{\infty} m(E \setminus K_n) < \frac{\epsilon}{2} + \frac{\epsilon}{2} = \epsilon.$$

Obviously, K_ϵ is closed in $\mathbb{R}$, every s_n is continuous and $s_n \to f$ uniformly on K_ϵ. Hence, by Theorem 10.6 (Uniform Convergence and Continuity), f is continuous on K_ϵ, completing the proof of the problem.   ∎

Remark 13.4

Traditionally, Theorem 13.9 (Lusin's Theorem) is proven by using Theorems 13.6 (The Simple Function Approximation Theorem) and 13.8 (Egorov's Theorem), just as what we have done in Problem 13.26. In fact, our argument used in Problems 13.25 and 13.26 follows that of Royden [17, pp. 66, 67]. Similar proofs can be found, for examples, in [19, Theorem 2.24, pp. 55, 56], [21, Theorem 4.5, p. 34] or [26, Theorem 5.29, pp. 137, 138]. In 2004, Loeb and Talvila [11] prove Theorem 13.9 (Lusin's Theorem) without using simple functions and their proof is presented in Problem 13.27.

Problem 13.27

⋆ ⋆ ⋆ *Prove Theorem 13.9 (Lusin's Theorem) with the closed set K_ϵ replaced by a compact set.*

Proof. Given a measurable set E with $m(E) < \infty$ and $\epsilon > 0$. We claim that there is a compact set $K \subseteq E$ such that

$$m(E \setminus K) < \epsilon.$$

To see this, we know from Remark 12.5 that there is a closed set F of $\mathbb{R}$ contained in E such that

$$m(E \setminus F) < \frac{\epsilon}{2}. \tag{13.34}$$

Since $m(E) < \infty$, we must have $m(F) < \infty$. Define $F_n = F \cap [-n, n]$ for every $n = 1, 2, \ldots$. It is evident that each F_n is measurable and compact. Furthermore, we have

$$F_1 \subseteq F_2 \subseteq \cdots \quad \text{and} \quad F = \bigcup_{n=1}^{\infty} F_n.$$

As a consequence of Theorem 12.12 (Continuity of Lebesgue Measure), we find that

$$\lim_{n \to \infty} m(F_n) = m(F).$$

Thus there exists an $N \in \mathbb{N}$ such that

$$m(F \setminus [-N, N]) = m(F \setminus F_N) < \frac{\epsilon}{2}. \tag{13.35}$$

Let $K = F_N$. Then it follows from the inequalities (13.34) and (13.35) that

$$m(E \setminus K) = m((E \setminus F) \cup (F \setminus K)) \leq m(E \setminus F) + m(F \setminus K) < \epsilon$$

which proves our claim.

Suppose that $\{V_n\}$ is an enumeration of the open intervals with rational endpoints in $\mathbb{R}$. It is clear that $\{V_n\}$ forms a basis of a topology of $\mathbb{R}$.[d] Fix an n. By Theorem 13.3, both $f^{-1}(V_n)$ and $E \setminus f^{-1}(V_n)$ are measurable. Since $f^{-1}(V_n)$ and $E \setminus f^{-1}(V_n)$ are subsets of E, they are of finite measures. Now the previous claim can be applied to conclude that there are compact sets $K_n \subseteq f^{-1}(V_n)$ and $K_n' \subseteq E \setminus f^{-1}(V_n)$ such that

$$m(E \setminus K_n) < \frac{\epsilon}{2^{n+1}} \quad \text{and} \quad m(E \setminus K_n') < \frac{\epsilon}{2^{n+1}}$$

and they imply that

$$m(E \setminus (K_n \cup K_n')) < \frac{\epsilon}{2^n}. \tag{13.36}$$

Set $K_\epsilon = \bigcap_{n=1}^{\infty} (K_n \cup K_n')$. Thus the estimate (13.36) give

$$m(E \setminus K_\epsilon) = m\left(\bigcup_{n=1}^{\infty} [E \setminus (K_n \cup K_n')] \right) \leq \sum_{n=1}^{\infty} m(E \setminus (K_n \cup K_n')) < \epsilon.$$

---

[d]See the definition on [12, p. 78].

This proves the first assertion.

To show that $f|_{K_\epsilon}$ is continuous, we recall from [12, Theorem 18.1, p. 104] that it is equivalent to showing that for each $x \in K_\epsilon$ and each neighborhood W of $f(x)$, there is a neighborhood U of x in E such that

$$f|_{K_\epsilon}(U) = f(K_\epsilon \cap U) \subseteq W. \qquad (13.37)$$

To verify this, given $x \in K_\epsilon$ and W a neighborhood of $f(x)$. We first notice that since $\{V_n\}$ is a basis of a topology of $\mathbb{R}$, there exists a basis element V_N such that

$$f(x) \in V_N \subseteq W.$$

Next, since $x \in f^{-1}(V_N)$ and $x \in K_\epsilon$, we gain $x \in K_N$. Let $U = E \setminus K'_N$. Since K'_N is compact, it is closed in E and then U is open in E. Recall the fact that $K_N \cap K'_N = \varnothing$, so we also have $x \in U$. Thus we establish

$$x \in U \cap K_\epsilon,$$

i.e., the set $U \cap K_\epsilon$ is a neighborhood of x in K_ϵ. Now for every $y \in U \cap K_\epsilon$, we have $y \in U$ so that $y \notin K'_N$. Since K_ϵ also contains y, the definition of K_ϵ implies that $y \in K_N$ and so

$$y \in f^{-1}(V_N),$$

i.e., $f(y) \in V_N \subseteq W$. Hence we conclude that

$$f|_{K_\epsilon}(U) = f(U \cap K_\epsilon) \subseteq V_N \subseteq W$$

which is exactly the set relation (13.37). This completes the proof of the problem. ∎

Problem 13.28

$(\star)(\star)$ *Suppose that E is measurable with $m(E) < \infty$ and $f : E \to \mathbb{R}$ is measurable. Given $\epsilon > 0$. Prove that there exists a step function $g : \mathbb{R} \to \mathbb{R}$ such that*

$$m(\{x \in E \mid |f(x) - g(x)| \geq \epsilon\}) < \epsilon.$$

Proof. By Problem 13.27, there exists a compact set $K \subseteq E$ such that

$$m(E \setminus K) < \epsilon \quad \text{and} \quad f|_K \text{ is continuous}.$$

As a compact set, K is bounded so that we can choose an $N \in \mathbb{N}$ such that K is a *proper* subset of $[-N, N]$ and $N \notin K$. Furthermore, $f|_K$ is uniformly continuous on K. Therefore, one can choose a $\delta \in (0, \epsilon)$ such that for all $p, q \in K$ with $|p - q| < \delta$, we have

$$|f(p) - f(q)| < \epsilon. \qquad (13.38)$$

Pick $n > \frac{1}{\delta}$. Let $x_k = -N + \frac{k}{n}$, where $k = 0, 1, 2, \ldots, 2nN$. Next, we denote

$$S = \{k \in \{0, 1, \ldots, 2nN - 1\} \mid [x_k, x_{k+1}) \cap K \neq \varnothing\}.$$

Now for each $k \in S$, we choose $p_k \in [x_k, x_{k+1}) \cap K$ and define $g : \mathbb{R} \to \mathbb{R}$ by

$$g(x) = \sum_{k \in S} f(p_k) \chi_{[x_k, x_{k+1})}(x).$$

which is a step function.

We observe that if $q \in K$, then there is a $t \in \{0, 1, \ldots, 2nN - 1\}$ such that $q \in [x_t, x_{t+1})$ and

$$|p_t - q| \le |x_{t+1} - x_t| = \frac{1}{n} < \delta.$$

Recall that $\chi_{[x_i, x_{i+1})}(q) = 1$ which implies that $g(q) = f(p_t)$. Using this fact and the inequality (13.38), we see that

$$|f(q) - g(q)| = |f(q) - f(p_t)| < \epsilon. \tag{13.39}$$

Since the inequality (13.39) holds for all $q \in K$, it implies that

$$m(\{x \in E \mid |f(x) - g(x)| \ge \epsilon\}) \le m(E \setminus K) < \epsilon$$

as required. This ends the proof of the problem. ∎

Problem 13.29

$(\star)(\star)$ *Suppose that E is measurable with $m(E) < \infty$ and $f : E \to \mathbb{R}$. Prove that there corresponds a F_σ set K and a sequence $\{f_n\}$ of continuous functions on K such that*

$$m(E \setminus K) = 0 \quad \text{and} \quad f_n \to f \text{ pointwise on } K.$$

Proof. For each $i \in \mathbb{N}$, we deduce from Theorem 13.9 (Lusin's Theorem) that there exists a closed set $K_i' \subseteq E$ such that

$$m(E \setminus K_i') < \frac{1}{i} \quad \text{and} \quad f \text{ is continuous on } K_i'. \tag{13.40}$$

For every $n \in \mathbb{N}$, we define

$$K_n = \bigcup_{i=1}^{n} K_i' \tag{13.41}$$

which is clearly a closed set in $\mathbb{R}$. Now our results (13.40) imply that f is continuous on K_n. Finally, we set

$$K = \bigcup_{n=1}^{\infty} K_n$$

which is a F_σ set by the definition. By the inequality (13.40), we know that

$$m(E \setminus K_1) = m(E \setminus K_1') < 1.$$

Besides, we have

$$\bigcap_{j=1}^{n+1} (E \setminus K_j) \subseteq \bigcap_{j=1}^{n} (E \setminus K_j)$$

for every $n = 1, 2, \ldots$, we find first from Theorem 12.12 (Continuity of Lebesgue Measure) and then using the definition (13.41) plus the inequality (13.40) to conclude that

$$m(E \setminus K) = m\left(E \setminus \bigcup_{n=1}^{\infty} K_n\right)$$

$$= m\Big(\bigcap_{n=1}^{\infty} (E \setminus K_n) \Big)$$

$$= \lim_{n\to\infty} m(E \setminus K_n)$$

$$= \lim_{n\to\infty} m\Big(\bigcap_{i=1}^{n} (E \setminus K_i') \Big)$$

$$\leq \lim_{n\to\infty} m(E \setminus K_n')$$

$$\leq \lim_{n\to\infty} \frac{1}{n}$$

and then $m(E \setminus K) = 0$.

Recall that f is continuous on the closed set K_n, so [18, Exercise 4.5, p. 99] implies that there exists a continuous function $f_n : \mathbb{R} \to \mathbb{R}$ such that

$$f_n(x) = f(x) \tag{13.42}$$

on K_n. By the definition (13.41), we have $K_n \subseteq K_{n'}$ if $n' \geq n$. Combining this observation and the result (13.42), we see that for every $x \in K$, there exists an $N \in \mathbb{N}$ such that $x \in K_N$ and therefore the equation (13.42) holds for all $n \geq N$. This gives the second assertion which completes the analysis of the problem. ∎

CHAPTER **14**

Lebesgue Integration

14.1 Fundamental Concepts

In Chapter 13, we have defined the notion of Lebesgue measurable functions. Now we are ready to define and study the theory of Lebesgue integration. The main references that we have applied are [5, Chap. 2], [9, Chap. 6], [17, Chap. 4, 5], [18, Chap. 11], [19, Chap. 1], [21, Chap. 2] and [22, Chap. 8].

14.1.1 Integration of Nonnegative Functions

Our starting point is the integration of a simple function.

Definition 14.1 (Integration of Simple Functions). *Suppose that $s : \mathbb{R} \to [0, \infty)$ is a simple function in the form*

$$s(x) = \sum_{k=1}^{n} c_k \chi_{E_k}(x),$$

*where $E_1, E_2, \ldots, E_n$ are measurable sets and $c_1, c_2, \ldots, c_k$ are distinct constants. Then the **Lebesgue integral** of s on a measurable set E is given by*

$$\int_{E} s \, dm = \sum_{k=1}^{n} c_k m(E_k \cap E).$$

Next, with the aid of Theorem 13.6 (The Simple Function Approximation Theorem), we have the integration of a nonnegative function.

Definition 14.2 (Integration of Nonnegative Functions). *Suppose that $f : \mathbb{R} \to [0, \infty]$ is measurable. Then the **Lebesgue integral** of f on a measurable set E is given by*

$$\int_{E} f \, dm = \sup \int_{E} s \, dm,$$

where the supremum is taken over all simple measurable functions s with $0 \leq s \leq f$.

The following theorem lists some useful and important properties of the Lebesgue integration of nonnegative functions.

Theorem 14.3. *Let* $f, g : \mathbb{R} \to [0, \infty]$ *and* $E, F, E_1, E_2, \ldots \subseteq \mathbb{R}$ *be measurable. Then we have*

(a) *If* $0 \le f \le g$, *then we have*

$$\int_E f \, dm \le \int_E g \, dm.$$

(b) *If* $a \le f(x) \le b$ *for all* $x \in E$, *then we have*

$$am(E) \le \int_E f \, dm \le bm(E).$$

(c) *If* $F \subseteq E$, *then we have*

$$\int_F f \, dm \le \int_E f \, dm.$$

(d) *If* $A, B \in [0, \infty)$, *then we have*

$$\int_E (Af + Bg) \, dm = A \int_E f \, dm + B \int_E g \, dm.$$

(e) *If* $E = \bigcup_{k=1}^{\infty} E_k$, *where* $E_1, E_2, \ldots$ *are pairwise disjoint, then we have*

$$\int_E f \, dm = \sum_{k=1}^{\infty} \int_{E_k} f \, dm.$$

(f) *If* $m(E) = 0$ *or* $f(x) = 0$ *for almost every* $x \in E$, *then we have*

$$\int_E f \, dm = 0.$$

(g) *If* $\int_E f \, dm = 0$ *and* $m(E) > 0$, *then we have* $f(x) = 0$ *for almost every* $x \in E$.

14.1.2 Integration of Lebesgue Integrable Functions

Definition 14.4 (Lebesgue Integrable Functions). *If* f *is a measurable function defined on the measurable set* $E \subseteq \mathbb{R}$ *and if*

$$\int_E |f| \, dm < \infty,$$

then f *is called a* **Lebesgue integrable function**. *The class of all Lebesgue integrable functions defined on* E *is denoted by* $L^1(E)$.

Remark 14.1

Recall from Remark 13.3 that the functions f^+ and f^- are nonnegative and $f = f^+ - f^-$. Since $f^\pm \leq |f|$, both functions f^+ and f^- are Lebesgue integrable provided that $f \in L^1(E)$ and then we can define the **Lebesgue integral** of f by

$$\int_E f \, \mathrm{d}m = \int_E f^+ \, \mathrm{d}m - \int_E f^- \, \mathrm{d}m.$$

Theorem 14.5 (Properties of Integrable Functions). *Suppose that E and F are measurable sets with $F \subseteq E$. Then we have the following properties:*

(a) *If $f, g \in L^1(E)$ and $A, B \in \mathbb{R}$, then we have $Af + Bg \in L^1(E)$ and*

$$\int_E (Af + Bg) \, \mathrm{d}m = A \int_E f \, \mathrm{d}m + B \int_E g \, \mathrm{d}m.$$

(b) *If $f, g \in L^1(E)$ and $f(x) \leq g(x)$ on E, then we have*

$$\int_E f \, \mathrm{d}m \leq \int_E g \, \mathrm{d}m.$$

(c) *If $f \in L^1(E)$, then we have $f \in L^1(F)$.*

(d) *If $f \in L^1(E)$, then $|f| \in L^1(E)$ and we have*

$$\left| \int_E f \, \mathrm{d}m \right| \leq \int_E |f| \, \mathrm{d}m.$$

If f is bounded by a positive constant M on the measurable set E with $m(E) < \infty$, then we have $|f| \leq M$ and Theorem 14.3(b) implies that

$$\int_E |f| \, \mathrm{d}m \leq M m(E) < \infty.$$

In this case, it is trivial that $f \in L^1(E)$ and the inequality in Theorem 14.5(d) (Properties of Integrable Functions) also holds.

▌ 14.1.3 Fatou's Lemma and Convergence Theorems

There are several elementary but remarkable results about integration of sequences of measurable functions.

Fatou's Lemma. *If each $f_n : \mathbb{R} \to [0, \infty]$ is measurable, then we have*

$$\int_E \left(\liminf_{n \to \infty} f_n \right) \mathrm{d}m \leq \liminf_{n \to \infty} \int_E f_n \, \mathrm{d}m.$$

The Bounded Convergence Theorem. *Let $\{f_n\}$ be a sequence of measurable functions defined on the measurable set E with $m(E) < \infty$. If $f_n \to f$ pointwise on E and $\{f_n\}$ is uniformly bounded[a], then we have*

$$\lim_{n \to \infty} \int_E f_n \, \mathrm{d}m = \int_E f \, \mathrm{d}m.$$

---

[a] See Definition 10.9 (Pointwise Boundedness and Uniformly Boundedness).

The Lebesgue's Monotone Convergence Theorem. *Suppose that E is measurable and $\{f_n\}$ is a sequence of measurable functions such that*

$$0 \le f_1(x) \le f_2(x) \le \cdots$$

on E. If $f_n(x) \to f(x)$ as $n \to \infty$ for almost all $x \in E$, then we have

$$\lim_{n \to \infty} \int_E f_n \, \mathrm{d}m = \int_E f \, \mathrm{d}m.$$

As an immediate application of the Lebesgue's Monotone Convergence Theorem, if every $f_n : \mathbb{R} \to [0, \infty]$ is measurable for $n = 1, 2, \ldots$ and

$$f(x) = \sum_{n=1}^{\infty} f_n(x)$$

exists a.e. on E, then we have

$$\int_E f \, \mathrm{d}m = \sum_{n=1}^{\infty} \int_E f_n \, \mathrm{d}m.$$

The Lebesgue's Dominated Convergence Theorem. *Suppose that E is measurable and $\{f_n\}$ is a sequence of measurable functions such that*

$$f(x) = \lim_{n \to \infty} f_n(x)$$

exists a.e. on E. If there exists a function $g \in L^1(E)$ such that

$$|f_n(x)| \le g(x)$$

for all $n \in \mathbb{N}$ and a.e. on E, then we have $f \in L^1(E)$ and

$$\lim_{n \to \infty} \int_E f_n \, \mathrm{d}m = \int_E f \, \mathrm{d}m.$$

The series version of the Lebesgue's Dominated Convergence Theorem is the following: suppose that $\{f_n\}$ is a sequence of measurable functions defined a.e. on E and

$$\sum_{n=1}^{\infty} \int_E |f_n| \, \mathrm{d}m < \infty.$$

Then the series

$$f(x) = \sum_{n=1}^{\infty} f_n(x)$$

converges for almost all $x \in E$, $f \in L^1(E)$ and

$$\int_E f \, \mathrm{d}m = \sum_{n=1}^{\infty} \int_E f_n \, \mathrm{d}m.$$

14.1.4 Connections between Riemann Integrals and Lebesgue Integrals

To distinguish Riemann integrals from Lebesgue integrals, we write the former by

$$\mathscr{R} \int_a^b f \, \mathrm{d}x.$$

Theorem 14.6. *Let $a, b \in \mathbb{R}$ and $a < b$.*

(a) *If $f \in \mathscr{R}$ on $[a, b]$, then $f \in L^1([a, b])$ and*

$$\int_{[a,b]} f \, \mathrm{d}m = \int_a^b f \, \mathrm{d}m = \mathscr{R} \int_a^b f \, \mathrm{d}x.$$

(b) *Let f be bounded on $[a, b]$. Then $f \in \mathscr{R}$ on $[a, b]$ if and only if f is continuous a.e. on $[a, b]$.*

14.2 Properties of Integrable Functions

Problem 14.1

$(\star)$ *For every $n \in \mathbb{N}$, calculate the Lebesgue integral*

$$\int_{\mathbb{R}} \frac{1}{n} \chi_{[0,n)} \, \mathrm{d}m.$$

Proof. Since $[0, n)$ is measurable, we obtain from Definition 14.1 (Integration of Simple Functions) directly that

$$\int_{\mathbb{R}} \frac{1}{n} \chi_{[0,n)} \, \mathrm{d}m = \frac{1}{n} m([0, n) \cap \mathbb{R}) = \frac{1}{n} m([0, n)) = 1$$

which completes the proof of the problem. ∎

Problem 14.2

$(\star)$ *Let $-\infty < a < b < \infty$ and consider the function $f : [a, b] \to \mathbb{R}$ defined by*

$$f(x) = \begin{cases} x, & \text{if } x \in [a, b] \setminus \mathbb{Q}; \\ 1, & \text{otherwise.} \end{cases}$$

Evaluate the Lebesgue integral of f on $[a, b]$.

Proof. Recall that $m(\mathbb{Q}) = 0$, so $m([a,b] \cap \mathbb{Q}) = 0$. Then we gain from Theorem 14.3(f) that

$$\int_{[a,b]} f \, dm = \int_{[a,b]} x \, dm.$$

Next, we conclude from [25, Theorem 9.4, p. 159] that $f \in \mathscr{R}$ on $[a,b]$. Thus it follows from Theorem 14.6(a) that $f \in L^1([a,b])$ and

$$\int_{[a,b]} f \, dm = \mathscr{R} \int_a^b x \, dx = \frac{b-a}{2},$$

completing the proof of the problem. ∎

Problem 14.3

$\bigstar$ *Prove that the collection $\mathscr{S}$ of simple functions defined on a set $E \subseteq \mathbb{R}$ is closed under addition.*

Proof. Let $x \in E$. Suppose that

$$s(x) = \sum_{i=1}^n a_i \chi_{A_i}(x) \quad \text{and} \quad t(x) = \sum_{j=1}^m b_j \chi_{B_j}(x), \tag{14.1}$$

where $A_1, A_2, \ldots, A_n$ and $B_1, B_2, \ldots, B_m$ are pairwise disjoint and $\bigcup_{i=1}^n A_i = \bigcup_{j=1}^m B_j = E$. Define $C_{ij} = A_i \cap B_j$ for $1 \le i \le n$ and $1 \le j \le m$. It is clear that

$$A_i = A_i \cap \bigcup_{j=1}^m B_j = \bigcup_{j=1}^m (A_i \cap B_j) = \bigcup_{j=1}^m C_{ij},$$

where $1 \le i \le n$. Similarly, we have

$$B_j = \bigcup_{j=1}^n C_{ij},$$

where $1 \le j \le m$. By the definition, it is true that

$$\bigcup_{i=1}^n \bigcup_{j=1}^m C_{ij} = E.$$

Furthermore, as $\{C_{ij}\}$ is a collection of pairwise disjoint sets, this means that

$$\chi_{A_i} = \sum_{j=1}^m \chi_{C_{ij}} \quad \text{and} \quad \chi_{B_j} = \sum_{i=1}^n \chi_{C_{ij}}. \tag{14.2}$$

By combining the expressions (14.1) and (14.2), we get

$$s(x) = \sum_{i=1}^n a_i \sum_{j=1}^m \chi_{C_{ij}}(x) \quad \text{and} \quad t(x) = \sum_{j=1}^m b_j \sum_{i=1}^n \chi_{C_{ij}}(x)$$

which imply that

$$s(x) + t(x) = \sum_{i=1}^{n} \sum_{j=1}^{m} (a_i + b_j) \chi_{C_{ij}}(x).$$

This completes the proof of the problem. ■

Remark 14.2

In fact, it can be shown further that the set $\mathscr{S}$ is also closed under multiplication.

Problem 14.4

$\circledast$ $\circledast$ *Let f be a bounded and nonnegative measurable function defined on $[0,1]$. Prove that*

$$\int_{[0,1]} f \, dm = \inf \int_{[0,1]} t \, dm, \qquad (14.3)$$

where the infimum is taken over all simple measurable functions such that $f \le t$.

Proof. Denote S and $u = \sup S$ to be a set and its supremum. Let $-S = \{-x \,|\, x \in S\}$. We claim that

$$\inf(-S) = -\sup S. \qquad (14.4)$$

On the one hand, we know from the definition that $x \le u$ for all $x \in S$ so that $-x \ge -u$. Thus $-u$ is a lower bound for the set $-S$, i.e.,

$$\inf(-S) \ge -u = -\sup S. \qquad (14.5)$$

On the other hand, let $v = \inf(-S)$. Then we have $v \le -x$ for all $x \in S$. Equivalently, $-v \ge x$ for all $x \in S$. Therefore, $-v$ is an upper bound of S and this implies that

$$-\inf(-S) = -v \ge \sup S. \qquad (14.6)$$

Combining the inequalities (14.5) and (14.6), we arrive at the claim (14.4).

Since f is bounded, there exists a $M > 0$ such that $0 \le f(x) \le M$ on $[0,1]$. Define $g : [0,1] \to \mathbb{R}$ by

$$g = M - f$$

which is a bounded (by $2M$) and nonnegative measurable function defined on $[0,1]$. Let t be a simple function such that $0 \le f \le t$. If we consider $t' = \min(t, M)$, then we have

$$0 \le t' \le M \quad \text{and} \quad f \le t' \le t.$$

Since $t' = \frac{1}{2}(t + M - |t - M|)$, Problem 14.3 implies that t' is simple. Note that

$$\int_{[0,1]} t' \, dm \le \int_{[0,1]} t \, dm$$

and so

$$\inf_{f \le t' \le M} \int_{[0,1]} t' \, dm = \inf_{f \le t} \int_{[0,1]} t \, dm.$$

Hence, in order to prove the representation (14.3), it suffices to consider simple functions t' such that $f \le t' \le M$.

Next, we know from Definition 14.2 (Integration of Nonnegative Functions) that

$$M - \int_{[0,1]} f \, dm = \int_{[0,1]} M \, dm - \int_{[0,1]} f \, dm = \int_{[0,1]} g \, dm = \sup \int_{[0,1]} s \, dm, \qquad (14.7)$$

where the supremum is taken over all simple measurable functions s with $0 \le s \le g$. Now $t = M - s$ is simple by Problem 14.3 and $f \le t \le M$. Conversely, if t is a simple function such that $f \le t \le M$, then $s = M - t$ is also a simple function with $0 \le s \le g$. Thus it follows from the formula (14.4) and the expression (14.7) that

$$\begin{aligned}
\int_{[0,1]} f \, dm &= \int_{[0,1]} (M - g) \, dm \\
&= M - \sup_{0 \le s \le g} \int_{[0,1]} s \, dm \\
&= \inf_{0 \le s \le g} \int_{[0,1]} M \, dm + \inf_{0 \le s \le g} \left(- \int_{[0,1]} s \, dm \right) \\
&= \inf_{0 \le s \le g} \int_{[0,1]} (M - s) \, dm \\
&= \inf_{f \le t \le M} \int_{[0,1]} t \, dm
\end{aligned}$$

which is our desired result. Hence, this completes the proof of the problem. ∎

Problem 14.5 (Chebyshev's Inequality)

(⋆) Suppose that f is a nonnegative and integrable function on a measurable set $E \subseteq \mathbb{R}$. If $\alpha > 0$, then prove that

$$m(\{x \in E \mid f(x) \ge \alpha\}) \le \frac{1}{\alpha} \int_E f \, dm. \qquad (14.8)$$

Proof. Let $F = \{x \in E \mid f(x) \ge \alpha\}$. Then F is measurable. If $m(F) = \infty$, then $m(E) = \infty$ and Theorem 14.3(b) implies that

$$\int_E f \, dm \ge \alpha \cdot m(E) = \infty$$

which contradicts Theorem 14.5(d) (Properties of Integrable Functions). In other words, we have $m(F) < \infty$. Now it is easy to see from Theorem 14.3(e) that

$$\int_E f \, dm = \int_F f \, dm + \int_{E \setminus F} f \, dm \ge \int_F f \, dm \ge \alpha \cdot m(F)$$

which is exactly the inequality (14.8). This ends the analysis of the problem. ∎

Problem 14.6

(⋆) Suppose that $f : E \to [0, \infty)$ is measurable, where E is a measurable set. If $f \in L^1(E)$, prove that f is finite a.e. on E.

Proof. Let $\alpha > 0$. Denote $F_\alpha = \{x \in E \mid f(x) \geq \alpha\}$ and $F_\infty = \{x \in E \mid f(x) = \infty\}$. Obviously, we have $F_\infty \subset F_n$ and $F_{n+1} \subseteq F_n$ for all $n \in \mathbb{N}$. In addition, we observe from Problem 14.5 and the integrability of f that

$$m(F_1) \leq \int_E f \, dm < \infty.$$

Hence Theorem 12.12 (Continuity of Lebesgue Measure) and Problem 14.5 imply that

$$m(F_\infty) \leq \lim_{n \to \infty} m(F_n) \leq \lim_{n \to \infty} \frac{1}{n} \int_E f \, dm = 0,$$

i.e., f is finite a.e. on E which ends the proof of the problem. ∎

Problem 14.7

($\star$) *Suppose that $E \subseteq \mathbb{R}$ is measurable and $f \in L^1(E)$ satisfies*

$$\left| \int_F f \, dm \right| < Cm(F)$$

for every measurable subset F of E with finite measure, where C is a positive constant. Prove that

$$|f(x)| < C$$

a.e. on E.

Proof. If $m(E) = 0$, then there is nothing to prove. Thus we assume that $m(E) > 0$. Since $f \in L^1(E)$, we have $|f| \in L^1(E)$ by Theorem 14.5(d) (Properties of Integrable Functions). Now Problem 14.5 guarantees that

$$m(\{x \in E \mid |f(x)| \geq C\}) \leq \frac{1}{C} \int_E |f| \, dm < \infty. \tag{14.9}$$

Consider the sets $F_1 = \{x \in E \mid f(x) \geq C\}$ and $F_2 = \{x \in E \mid f(x) \leq -C\}$. By the estimate (14.9), both $m(F_1)$ and $m(F_2)$ are finite.

On the set F_1, the hypothesis implies that

$$Cm(F_1) \leq \int_{F_1} f \, dm = \left| \int_{F_1} f \, dm \right| < Cm(F_1)$$

so that $m(F_1) = 0$. On the set F_2, the hypothesis gives

$$Cm(F_2) \leq \left| \int_{F_2} f \, dm \right| < Cm(F_2)$$

which shows again that $m(F_2) = 0$. Equivalently, we conclude that $|f(x)| < C$ a.e. on E and this completes the proof of the problem. ∎

Problem 14.8

($\star$) *Suppose that E and f are measurable and $\int_E |f| \, dm = 0$. Prove that $f(x) = 0$ a.e. on E.*

Proof. For every $n = 1, 2, \ldots$, we let $E_n = \{x \in E \mid |f(x)| \geq \frac{1}{n}\}$. On the one hand, we have

$$\{x \in E \mid f(x) \neq 0\} = \bigcup_{n=1}^{\infty} E_n,$$

so we deduce from the inequality (12.1) that

$$m(\{x \in E \mid f(x) \neq 0\}) \leq \sum_{n=1}^{\infty} m(E_n). \tag{14.10}$$

On the other hand, we know from Problem 14.5 that

$$m(E_n) \leq n \int_E |f| \, dm = 0. \tag{14.11}$$

Hence we conclude from the inequalities (14.10) and (14.11) that

$$m(\{x \in E \mid f(x) \neq 0\}) = 0$$

which is equivalent to the condition $f(x) = 0$ a.e. on E, completing the proof of the problem. ∎

Problem 14.9

⋆ ⋆ *Let $f : [0,1] \to \mathbb{R}$ be a measurable function. Prove that $f \in L^1([0,1])$ if and only if*

$$\sum_{n=1}^{\infty} 2^n \cdot m(\{x \in [0,1] \mid |f(x)| \geq 2^n\}) < \infty. \tag{14.12}$$

Proof. Define $E_0 = \{x \in [0,1] \mid |f(x)| < 2\}$ and $g(x) = 1$ on E_0. Next, for each $n \in \mathbb{N}$, we define

$$E_n = \{x \in [0,1] \mid 2^n \leq |f(x)| < 2^{n+1}\} \quad \text{and} \quad g(x) = 2^n \text{ on } E_n. \tag{14.13}$$

It is easy to check that

$$m(E_0) \leq 1 \quad \text{and} \quad m(E_n) \leq m(\{x \in [0,1] \mid |f(x)| \geq 2^n\}) \tag{14.14}$$

for all $n \in \mathbb{N}$. Furthermore, it is also true that $0 \leq g(x) - 1 < |f(x)| \leq 2g(x)$ for all $x \in [0,1]$, thus we get from Theorem 14.3(a) that

$$\int_{[0,1]} (g-1) \, dm \leq \int_{[0,1]} |f| \, dm < 2 \int_{[0,1]} g \, dm$$

and these mean that $f \in L^1([0,1])$ if and only if $g \in L^1([0,1])$.

We claim that $g \in L^1([0,1])$ if and only if the condition (14.12) holds. To see this, we note from the definition (14.13) that $E_1, E_2, \ldots$ are pairwise disjoint and

$$\{x \in [0,1] \mid |f(x)| \geq 2^n\} = E_n \cup E_{n+1} \cup \cdots,$$

where $n = 1, 2, \ldots$. Therefore, we follow from Theorem 12.11 (Countably Additivity) that

$$m(\{x \in [0,1] \mid |f(x)| \geq 2^n\}) = \sum_{k=n}^{\infty} m(E_k), \tag{14.15}$$

where $n = 1, 2, \ldots$. Now we establish from Theorem 14.3(e), the relations (14.14) and finally the hypothesis (14.12) that

$$
\begin{aligned}
\int_{[0,1]} g \, dm &= \sum_{n=0}^{\infty} \int_{E_n} g \, dm \\
&= \sum_{n=0}^{\infty} 2^n \cdot m(E_n) \\
&\leq 1 + \sum_{n=1}^{\infty} 2^n \cdot m(\{x \in [0,1] \,|\, |f(x)| \geq 2^n\}).
\end{aligned}
\tag{14.16}
$$

On the other hand, we observe from the expression (14.15) that

$$
\begin{aligned}
\sum_{n=1}^{\infty} 2^n \cdot m(\{x \in [0,1] \,|\, |f(x)| \geq 2^n\}) &= \sum_{n=1}^{\infty} 2^n \cdot \sum_{k=n}^{\infty} m(E_k) \\
&= \sum_{k=1}^{\infty} m(E_k) \sum_{n=1}^{k} 2^n \\
&\leq \sum_{k=1}^{\infty} 2^{k+1} m(E_k) \\
&\leq 2 \sum_{k=0}^{\infty} 2^k m(E_k) \\
&= 2 \int_{[0,1]} g \, dm.
\end{aligned}
\tag{14.17}
$$

By the inequalities (14.16) and (14.17), we see that

$$
\frac{1}{2} \sum_{n=1}^{\infty} 2^n \cdot m(\{x \in [0,1] \,|\, |f(x)| \geq 2^n\}) \leq \int_{[0,1]} g \, dm \leq 1 + \sum_{n=1}^{\infty} 2^n \cdot m(\{x \in [0,1] \,|\, |f(x)| \geq 2^n\})
$$

which implies that $g \in L^1([0,1])$ if and only if the hypothesis (14.12) holds. This completes the proof of the problem. ∎

Problem 14.10

⋆ ⋆ Let E be measurable with $m(E) < \infty$ and f be nonnegative and integrable on E. Given $\epsilon > 0$, define

$$
E_n = \{x \in E \,|\, n\epsilon \leq f(x) < (n+1)\epsilon\}
$$

for every $n = 0, 1, 2, \ldots$ and

$$
A(\epsilon) = \epsilon \sum_{n=0}^{\infty} n \cdot m(E_n).
$$

Prove that

$$
\int_E f \, dm = \lim_{\epsilon \to 0} A(\epsilon).
$$

Proof. Define $F = \{x \in E \mid f(x) = \infty\}$. By the definition, we know that $E_1, E_2, \ldots$ are pairwise disjoint and

$$E = F \cup \bigcup_{n=0}^{\infty} E_n.$$

On the one hand, we obtain from Theorem 14.3(e) and Problem 14.6 that

$$A(\epsilon) = \epsilon \sum_{n=0}^{\infty} n \cdot m(E_n) = \sum_{n=0}^{\infty} \int_{E_n} (n\epsilon) \, dm \le \sum_{n=0}^{\infty} \int_{E_n} f \, dm = \int_{E \setminus F} f \, dm = \int_{E} f \, dm. \quad (14.18)$$

On the other hand, we see from Theorem 14.3(e) and Problem 14.6 again that

$$\begin{aligned}
A(\epsilon) + \epsilon m(E) &= \epsilon \sum_{n=0}^{\infty} n \cdot m(E_n) + \epsilon \cdot \sum_{n=0}^{\infty} m(E_n) \\
&= \sum_{n=0}^{\infty} (n+1)\epsilon m(E_n) \\
&= \sum_{n=0}^{\infty} \int_{E_n} (n+1)\epsilon \, dm \\
&> \sum_{n=0}^{\infty} \int_{E_n} f \, dm \\
&= \int_{E \setminus F} f \, dm \\
&= \int_{E} f \, dm. \quad (14.19)
\end{aligned}$$

Hence we deduce from the inequalities (14.18) and (14.19) that

$$\int_{E} f \, dm - \epsilon m(E) < A(\epsilon) \le \int_{E} f \, dm$$

which implies the desired result by letting $\epsilon \to 0$. This completes the proof of the problem. ■

Problem 14.11

(★) *Let E be a measurable set and f be measurable on E. If there is a nonnegative function $g \in L^1(E)$ such that*

$$|f(x)| \le g(x)$$

on E, prove that $f \in L^1(E)$.

Proof. By Theorem 14.3(a) and the hypothesis, we have

$$\int_{E} |f| \, dm \le \int_{E} g \, dm < \infty,$$

i.e., $f \in L^1(E)$ which completes the proof of the problem. ■

Problem 14.12

$(\star)$ Let $E \in \mathscr{B}$ and f be integrable on E. Define

$$\varphi(E) = \int_E f \, dm$$

for every $E \in \mathscr{B}$. Prove that $\varphi(E) = 0$ for all $E \in \mathscr{B}$ if and only if $f = 0$ a.e. on $\mathbb{R}$.

Proof. If $f = 0$ a.e. on $\mathbb{R}$, then it follows form Theorem 14.3(f) that

$$\varphi(E) = \int_E f \, dm = 0$$

for every $E \in \mathscr{B}$. Conversely, consider $E = \{x \in \mathbb{R} \mid f(x) \geq 0\}$. Then we have $f^+ = f \cdot \chi_E$ so that

$$\int_{\mathbb{R}} f^+ \, dm = \int_E f \, dm = \varphi(E) = 0.$$

In other words, we have $f^+ = 0$ a.e. on $\mathbb{R}$ by Theorem 14.3(g). Similarly, we can show that

$$\int_{\mathbb{R}} f^- \, dm = 0$$

so that $f^- = 0$ a.e. on $\mathbb{R}$. By the definition, $f = f^+ - f^- = 0$ a.e. on $\mathbb{R}$ which completes the proof of the problem. ∎

Problem 14.13

$(\star)$ Suppose that E is measurable and $f, g, h : E \to \mathbb{R}^*$ are measurable. If g and h are integrable on E and $g \leq f \leq h$ a.e. on E, prove that f is integrable on E.

Proof. Since h and g are integrable on E, $h - g$ is integrable on E by Theorem 14.5(a) (Properties of Integrable Functions). Obviously, we have

$$0 \leq f - g \leq h - g$$

a.e. on E, so Problem 14.11 implies that $f - g$ is also integrable on E. By Theorem 14.5(a) (Properties of Integrable Functions) again, the function

$$f = (f - g) + g$$

is integrable on E. This completes the proof of the problem. ∎

Problem 14.14

$(\star)$ Suppose that E is measurable, $E_1, E_2, \ldots$ are measurable subsets of E and $f : E \to \mathbb{R}^*$ is a measurable function. Let $F = \bigcup\limits_{n=1}^{\infty} E_n$. If f is integrable on each E_n and $\sum\limits_{n=1}^{\infty} \int_{E_n} |f| \, dm$ converges, prove that f is integrable on F and

$$\left| \int_F f \, dm \right| \leq \sum_{n=1}^{\infty} \int_{E_n} |f| \, dm.$$

Proof. Without loss of generality, we may assume that $E_1, E_2, \ldots$ are pairwise disjoint.[b] Recall from Remark 14.1 that $|f^{\pm}| \leq |f|$ on each E_n, so we have

$$\int_{E_n} f^{\pm} \, dm \leq \int_{E_n} |f| \, dm \tag{14.20}$$

for every $n = 1, 2, \ldots$. Since $f^{\pm} \geq 0$ on every E_n, we obtain from Theorem 14.3(e), the estimate (14.20) and then the hypothesis that

$$\int_F f^{\pm} \, dm = \sum_{n=1}^{\infty} \int_{E_n} f^{\pm} \, dm \leq \sum_{n=1}^{\infty} \int_{E_n} |f| \, dm < \infty. \tag{14.21}$$

Recall that $|f| = f^+ + f^-$, so the estimates (14.21) give

$$\int_F |f| \, dm = \int_F f^+ \, dm + \int_F f^- \, dm < \infty.$$

By Definition 14.4 (Lebesgue Integrable Functions), the function f is integrable on F which proves the firs assertion.

For the second assertion, we notice from Theorem 14.5(d) (Properties of Integrable Functions) and then Theorem 14.3(e) that

$$\left| \int_F f \, dm \right| \leq \int_F |f| \, dm = \sum_{n=1}^{\infty} \int_{E_n} |f| \, dm.$$

This completes the analysis of the problem.   ∎

Problem 14.15

$(\star)(\star)$ Suppose that E is measurable with $m(E) < \infty$ and $f, g : E \to \mathbb{R}$ are measurable. If $f, g \in L^1(E)$ and

$$\int_F f \, dm \leq \int_F g \, dm$$

for every measurable set $F \subseteq E$. Verify that $f(x) \leq g(x)$ a.e. on E.

---

[b]See the proof of Problem 12.25.

Proof. Consider

$$E_n = \left\{ x \in E \,\middle|\, f(x) \geq g(x) + \frac{1}{n} \right\},$$

where $n = 1, 2, \ldots$. Then we follow from Problems 13.6 and 13.8 that each

$$E_n = \left\{ x \in E \,\middle|\, f(x) > g(x) + \frac{1}{n} \right\} \cap \left\{ x \in E \,\middle|\, f(x) - g(x) = \frac{1}{n} \right\} \tag{14.22}$$

is measurable. Given $n \in \mathbb{N}$ fixed, since $g \in L^1(E)$, Theorem 14.5(c) and (d) (Properties of Integrable Functions) say that

$$\left| \int_{E_n} g \, dm \right| \leq \int_{E_n} |g| \, dm < \infty. \tag{14.23}$$

Furthermore, we see immediately from the hypothesis and the expression (14.22) that

$$\int_{E_n} g \, dm \geq \int_{E_n} f \, dm \geq \int_{E_n} g \, dm + \frac{1}{n} m(E_n) \tag{14.24}$$

for every $n \in \mathbb{N}$. Consequently, we conclude from the inequalities (14.23) and (14.24) that $m(E_n) = 0$ for each $n = 1, 2, \ldots$. Thus we get from the inequality (12.1) that

$$m \left(\bigcup_{n=1}^{\infty} E_n \right) = 0. \tag{14.25}$$

Let $E' = \{ x \in E \,|\, f(x) > g(x) \}$. If $p \in E'$, then we must have $f(p) > g(p) + \frac{1}{n}$ for some large enough $n \in \mathbb{N}$, i.e., $p \in E_n$. This means, with the aid of the result (14.25), that

$$m(E') = 0$$

and this is equivalent to $f(x) \leq g(x)$ a.e. on E. We have completed the proof of the problem. ∎

Problem 14.16

⭐⭐ *Suppose that $f : [0,1] \to [0, \infty)$ is measurable. Consider $0 < \epsilon \leq 1$ and*

$$S = \left\{ \int_E f \, dm \,\middle|\, E \text{ is a measurable subset of } [0,1] \text{ with } m(E) \geq \epsilon \right\}.$$

Prove that $\inf(S) > 0$.

Proof. Define $E_0 = \{ x \in [0,1] \,|\, f(x) \geq 1 \}$ and for each $n \in \mathbb{N}$, we define

$$E_n = \left\{ x \in [0,1] \,\middle|\, \frac{1}{n+1} \leq f(x) < \frac{1}{n} \right\}.$$

Then it is trivial that

$$[0,1] = E_0 \cup \bigcup_{n=1}^{\infty} E_n. \tag{14.26}$$

Since $E_0, E_1, E_2, \ldots$ are pairwise disjoint measurable subsets of $[0,1]$, we apply Theorem 12.11 (Countably Additivity) to the representation (14.26) to get

$$\sum_{n=0}^{\infty} m(E_n) = 1$$

which implies the existence of an $N \in \mathbb{N}$ such that

$$\sum_{n=N}^{\infty} m(E_n) \leq \frac{\epsilon}{2}. \tag{14.27}$$

By Problem 12.27, there exists a measurable subset $F \subseteq [0,1]$ such that $m(F) = \epsilon$. Let

$$F_1 = F \cap \left\{ x \in [0,1] \,\middle|\, f(x) \geq \frac{1}{N+1} \right\} = F \cap \bigcup_{n=N}^{\infty} E_n \quad \text{and} \quad F_2 = F \setminus F_1.$$

Then we see from Problem 12.13 and the estimate (14.27) that

$$m(F_2) = m(F \setminus F_1) = m(F) - m(F_1) \geq \epsilon - \sum_{n=N}^{\infty} m(E_n) \geq \frac{\epsilon}{2} > 0.$$

Hence it yields from Theorem 14.3(a) that

$$\int_F f \, dm \geq \int_{F_2} f \, dm \geq \frac{1}{N+1} m(F_2) \geq \frac{\epsilon}{2(N+1)} > 0.$$

In other words, we have $\inf(S) > 0$ which completes the proof of the problem. ∎

Problem 14.17

⋆ *Is it true that if $f, g : \mathbb{R} \to \mathbb{R}$ are integrable, then $f \circ g : \mathbb{R} \to \mathbb{R}$ is integrable?*

Proof. The answer is negative. Consider $f = \chi_{[0,1]}$ and $g = \chi_{\{1\}}$. Clearly, we have $f, g \in L^1(\mathbb{R})$. However, we note that

$$
\begin{aligned}
f(g(x)) &= \chi_{[0,1]}(\chi_{\{1\}}(x)) \\
&= \begin{cases} \chi_{[0,1]}(1), & \text{if } x = 1; \\[2mm] \chi_{[0,1]}(0), & \text{otherwise} \end{cases} \\
&= 1
\end{aligned}
$$

which implies that $f \circ g \notin L^1(\mathbb{R})$ because

$$\int_{\mathbb{R}} |f(g(x))| \, dm = m(\mathbb{R}) = \infty.$$

This completes the proof of the problem. ∎

Problem 14.18

$(\star)$ *Suppose that E is measurable with $m(E) > 0$ and f is measurable on E with*

$$\left| \int_E f \, dm \right| = \int_E |f| \, dm.$$

Prove that either $f \geq 0$ a.e. on E or $f \leq 0$ a.e. on E.

Proof. Suppose that

$$\int_E f \, dm \geq 0. \tag{14.28}$$

By the hypothesis, we have

$$\int_E (f - |f|) \, dm = 0,$$

so Theorem 14.3(g) implies that $f(x) - |f(x)| = 0$ a.e. on E and this means that

$$f(x) = |f(x)| \geq 0$$

a.e. on E.

Next, we suppose that

$$\int_E f \, dm \leq 0.$$

In this case, we consider the function $-f$ which is also measurable on E and it satisfies condition (14.28). Thus our previous analysis can be applied to $-f$ to conclude that

$$-f(x) = |-f(x)| \geq 0$$

a.e. on E and this is equivalent to

$$f(x) \leq 0$$

a.e. on E. This completes the proof of the problem. ∎

Problem 14.19

$(\star)$ *Suppose that $f \in L^1((0, \infty))$. Prove that there exists a sequence $x_n \to \infty$ such that*

$$\lim_{n \to \infty} x_n f(x_n) = 0.$$

Proof. We put

$$\alpha = \liminf_{x \to \infty} x|f(x)|.$$

If $\alpha = 0$, then since $-x|f(x)| \leq xf(x) \leq x|f(x)|$ on $(0, \infty)$, we have

$$\liminf_{x \to \infty} xf(x) = 0$$

and, by the definition, such a sequence exists. Next, we suppose that $\alpha > 0$. Then there is an $N \in \mathbb{N}$ such that $x|f(x)| > \frac{\alpha}{2}$ for all $x \geq N$. Consequently, we conclude from Theorem 14.3(c) that
$$\int_{(0,\infty)} |f| \, dm \geq \int_{(N,\infty)} |f| \, dm \geq \int_{N}^{\infty} \frac{\alpha}{2x} \, dx = \frac{\alpha}{2} \ln x \big|_N^\infty = \infty,$$
a contradiction. Hence we have completed the proof of the problem. ■

Problem 14.20

$\bigstar$ $\bigstar$ *Suppose that E is measurable and f is integrable on E. Given $\epsilon > 0$. Prove that there is a $\delta > 0$ such that*
$$\left| \int_F f \, dm \right| < \epsilon \tag{14.29}$$
for every $F \subseteq E$ with $m(F) < \delta$.

Proof. Given $\epsilon > 0$. If f is bounded on E, then there exists a $M > 0$ such that $|f(x)| \leq M$. Since $f \in L^1(E)$, Theorem 14.5(d) (Properties of Integrable Functions) gives
$$\left| \int_F f \, dm \right| \leq M \cdot m(F). \tag{14.30}$$
If F is a measurable subset of E such that $m(F) < \frac{\epsilon}{M}$, then we conclude from the inequality (14.30) that the inequality (14.29) holds in this case.

Suppose next that f is unbounded on E. For each $n = 0, 1, 2, \ldots$, we consider the sets
$$E_n = \{x \in E \mid n \leq |f(x)| < n+1\}, \quad F_n = \bigcup_{k=0}^{n} E_k \quad \text{and} \quad G_n = E \setminus F_n = \bigcup_{k=N+1}^{n} E_k. \tag{14.31}$$
Since $|f(x)| \geq 0$ on E and $E_0, E_1, \ldots$ are pairwise disjoint subsets of E whose union is E, we know from Theorem 14.3(e) and the hypothesis that
$$\sum_{n=0}^{\infty} \int_{E_n} |f| \, dm = \int_E |f| \, dm < \infty.$$
Therefore, there exists an $N \in \mathbb{N}$ such that
$$\int_{G_N} |f| \, dm = \sum_{n=N+1}^{\infty} \int_{E_n} |f| \, dm < \frac{\epsilon}{2}. \tag{14.32}$$
Let $\delta = \frac{\epsilon}{2(N+1)}$. Thus if F is a measurable subset of E with $m(F) < \delta$, then it yields from Theorem 14.5(d) (Properties of Integrable Functions), the facts $E = F_N \cup G_N$ and $F_N \cap G_N = \varnothing$ that
$$\left| \int_F f \, dm \right| \leq \int_F |f| \, dm = \int_{F \cap F_N} |f| \, dm + \int_{F \cap G_N} |f| \, dm. \tag{14.33}$$
By applying the definition (14.31) and the inequality (14.32) to the expression (14.33), we obtain immediately that
$$\left| \int_F f \, dm \right| \leq (N+1)m(F) + \int_{G_N} |f| \, dm < (N+1)\delta + \frac{\epsilon}{2} = \frac{\epsilon}{2} + \frac{\epsilon}{2} = \epsilon.$$
This completes the proof of the problem. ■

14.3 Applications of Fatou's Lemma

Problem 14.21

$(\star)$ Prove that we may have the strict inequality in Fatou's Lemma.

Proof. For each $n = 1, 2, \ldots$, we consider the function $f_n : \mathbb{R} \to \mathbb{R}$ defined by

$$f_n(x) = \begin{cases} 1, & \text{if } x \in [n, n+1); \\ 0, & \text{otherwise.} \end{cases} \tag{14.34}$$

If $x \in \mathbb{R}$, then it is easy to check that $x \notin [n, n+1)$ for all large enough n so that

$$\liminf_{n \to \infty} f_n(x) = 0$$

which gives

$$\int_{\mathbb{R}} \Big(\liminf_{n \to \infty} f_n \Big) \, dm = 0.$$

However, the definition of f_n implies that

$$\int_{\mathbb{R}} f_n \, dm = \int_n^{n+1} dm = 1.$$

Hence the sequence of functions (14.34) shows that the strict inequality in Fatou's Lemma can actually occur, completing the proof of the problem. ∎

Problem 14.22

$(\star)(\star)$ Suppose that $f_n, g_n, f, g \in L^1(\mathbb{R})$ for every $n \in \mathbb{N}$. If $f_n \to f$, $g_n \to g$ pointwise a.e. on $\mathbb{R}$, $|f_n(x)| \le g_n(x)$ a.e. on $\mathbb{R}$ and

$$\int_{\mathbb{R}} g_n \, dm \to \int_{\mathbb{R}} g \, dm,$$

prove that

$$\int_{\mathbb{R}} f_n \, dm \to \int_{\mathbb{R}} f \, dm.$$

Proof. By the triangle inequality, we know that $|f_n - f| \le |f_n| + |f| \le g_n + |f|$ so that

$$g_n + |f| - |f_n - f| \ge 0.$$

By Fatou's Lemma, we have

$$\int_{\mathbb{R}} g \, dm + \int_{\mathbb{R}} |f| \, dm = \int_{\mathbb{R}} (g + |f|) \, dm$$

$$= \int_{\mathbb{R}} \lim_{n \to \infty} (g + |f| - |f_n - f|) \, dm$$

$$= \int_{\mathbb{R}} \liminf_{n\to\infty} (g + |f| - |f_n - f|)\, dm$$

$$\leq \liminf_{n\to\infty} \int_{\mathbb{R}} (g + |f| - |f_n - f|)\, dm$$

$$= \liminf_{n\to\infty} \left(\int_{\mathbb{R}} g\, dm + \int_{\mathbb{R}} |f|\, dm - \int_{\mathbb{R}} |f_n - f|\, dm \right)$$

$$= \int_{\mathbb{R}} g\, dm + \int_{\mathbb{R}} |f|\, dm - \limsup_{n\to\infty} \int_{\mathbb{R}} |f_n - f|)\, dm$$

so that

$$\limsup_{n\to\infty} \int_{\mathbb{R}} |f_n - f|)\, dm = 0. \tag{14.35}$$

Since $f_n, f \in L^1(E)$, we deduce from Theorem 14.5(a) and (d) (Properties of Integrable Functions) that $f_n - f \in L^1(E)$ and

$$\left| \int_{\mathbb{R}} f_n\, dm - \int_{\mathbb{R}} f\, dm \right| = \left| \int_{\mathbb{R}} (f_n - f)\, dm \right| \leq \int_{\mathbb{R}} |f_n - f|\, dm. \tag{14.36}$$

Applying the result (14.35) to the inequality (14.36), we see that

$$\left| \int_{\mathbb{R}} f_n\, dm - \int_{\mathbb{R}} f\, dm \right| \to 0$$

as $n \to \infty$ which implies that

$$\lim_{n\to\infty} \int_{\mathbb{R}} f_n\, dm = \int_{\mathbb{R}} f\, dm.$$

This ends the proof of the problem. ∎

Problem 14.23

(⋆) Let $-\infty < a < b < \infty$. Let $\{f_n\}$ be a sequence of nonnegative measurable functions on (a, b) such that $f_n \to f$ a.e. on (a, b). Define

$$F(x) = \int_a^x f\, dm \quad \text{and} \quad F_n(x) = \int_a^x f_n\, dm,$$

where $n = 1, 2, \ldots$. Prove that

$$\int_{[a,b]} (f + F)\, dm \leq \liminf_{n\to\infty} \int_{[a,b]} (f_n + F_n)\, dm. \tag{14.37}$$

Proof. Since $\{f_n\}$ satisfies the hypothesis of Fatou's Lemma, so we know that

$$F(x) = \int_a^x f\, dm = \int_a^x \liminf_{n\to\infty} f_n\, dm \leq \liminf_{n\to\infty} \int_a^x f_n\, dm = \liminf_{n\to\infty} F_n(x). \tag{14.38}$$

Since f_n is nonnegative, each F_n is also nonnegative. Now we apply Fatou's Lemma to the inequality (14.38) one more time, we see that

$$\int_{[a,b]} (f + F)\, dm \leq \int_{[a,b]} (f + \liminf_{n\to\infty} F_n)\, dm$$

$$= \int_{[a,b]} \liminf_{n\to\infty}(f_n + F_n)\,\mathrm{d}m$$

$$\leq \liminf_{n\to\infty} \int_{[a,b]} (f_n + F_n)\,\mathrm{d}m$$

which is exactly the expected result (14.37). It completes the proof of the problem. ∎

Problem 14.24

$(\star)$ Suppose that E is measurable and $\{f_n\}$ is a sequence of measurable functions defined on E. If g is integrable on E and $|f_n| \leq g$ on E for all $n = 1, 2, \ldots$, prove that

$$\int_E \liminf f_n\,\mathrm{d}m \leq \liminf_{n\to\infty} \int_E f_n\,\mathrm{d}m \leq \limsup_{n\to\infty} \int_E f_n\,\mathrm{d}m \leq \int_E \limsup f_n\,\mathrm{d}m. \qquad (14.39)$$

Proof. Since $|f_n| \leq g$ on E and for all $n = 1, 2, \ldots$, we must have $g - f_n \geq 0$ and $g + f_n \geq 0$ on E and for all $n = 1, 2, \ldots$. Applying Fatou's Lemma to both $g - f_n$ and $g + f_n$, we get

$$\int_E \liminf_{n\to\infty}(g + f_n)\,\mathrm{d}m \leq \liminf_{n\to\infty} \int_E (g + f_n)\,\mathrm{d}m$$

and

$$\int_E \liminf_{n\to\infty}(g - f_n)\,\mathrm{d}m \leq \liminf_{n\to\infty} \int_E (g - f_n)\,\mathrm{d}m.$$

Then they imply that

$$\int_E \liminf_{n\to\infty} f_n\,\mathrm{d}m + \int_E g\,\mathrm{d}m \leq \liminf_{n\to\infty} \int_E f_n\,\mathrm{d}m + \int_E g\,\mathrm{d}m \qquad (14.40)$$

and

$$\int_E g\,\mathrm{d}m - \int_E \limsup_{n\to\infty} f_n\,\mathrm{d}m \leq \int_E g\,\mathrm{d}m - \limsup_{n\to\infty} \int_E f_n\,\mathrm{d}m. \qquad (14.41)$$

Combining the results (14.40) and (14.41), we conclude immediately that the set of inequalities (14.39) hold and we complete the proof of the problem. ∎

Problem 14.25

$(\star)(\star)$ Suppose that E is a measurable set and $\{f_n\}$ is a sequence of integrable functions defined on E that converges pointwise almost everywhere to an integrable function f defined on E. If we have

$$\lim_{n\to\infty} \int_E |f_n|\,\mathrm{d}m = \int_E |f|\,\mathrm{d}m, \qquad (14.42)$$

prove that

$$\lim_{n\to\infty} \int_F |f_n|\,\mathrm{d}m = \int_F |f|\,\mathrm{d}m$$

for every measurable subset F of E.

Proof. Let F be a measurable subset of E. Notice that

$$|f_n - f|\chi_F \le |f_n - f| \le |f_n| + |f|$$

on E. Thus if we set $g_n = |f_n| + |f| - |f_n - f|\chi_F$, then they are nonnegative and measurable so that we can apply Fatou's Lemma to get

$$\int_E \liminf_{n\to\infty} g_n \, dm \le \liminf_{n\to\infty} \int_E g_n \, dm$$

$$\int_E \liminf_{n\to\infty}(|f_n| + |f| - |f_n - f|\chi_F) \, dm \le \liminf_{n\to\infty} \int_E (|f_n| + |f| - |f_n - f|\chi_F) \, dm. \qquad (14.43)$$

Since $f_n \to f$ pointwise a.e. on E, we deduce from the hypothesis (14.42) and the inequality (14.43) that

$$2\int_E |f| \, dm - \int_E \limsup_{n\to\infty} |f_n - f|\chi_F \, dm \le \liminf_{n\to\infty} \int_E |f_n| \, dm + \int_E |f| \, dm$$
$$- \limsup_{n\to\infty} \int_E |f_n - f|\chi_F \, dm$$

$$2\int_E |f| \, dm - 0 \le \int_E |f| \, dm + \int_E |f| \, dm - \limsup_{n\to\infty} \int_F |f_n - f| \, dm$$

$$\limsup_{n\to\infty} \int_F |f_n - f| \, dm \le 0$$

which implies

$$0 \le \liminf_{n\to\infty} \int_F |f_n - f| \, dm \le \limsup_{n\to\infty} \int_F |f_n - f| \, dm \le 0. \qquad (14.44)$$

Consequently, the inequalities (14.44) mean that

$$\lim_{n\to\infty} \int_F |f_n - f| \, dm = 0. \qquad (14.45)$$

By Theorem 14.5(a) and (d) (Properties of Integrable Functions) and the triangle inequality, we obtain

$$\left| \int_F |f_n| \, dm - \int_F |f| \, dm \right| = \left| \int_F (|f_n| - |f|) \, dm \right| \le \int_F ||f_n| - |f|| \, dm \le \int_F |f_n - f| \, dm.$$

Therefore, the limit (14.45) implies that

$$\lim_{n\to\infty} \left| \int_F |f_n| \, dm - \int_F |f| \, dm \right| = 0$$

which gives the desired result. This completes the proof of the problem. $\blacksquare$

Problem 14.26

$(\star)(\star)$ *Use Fatou's Lemma to prove the Lebesgue's Monotone Convergence Theorem.*

Proof. Since $f_n \to f$ pointwise a.e. on E and every f_n is nonnegative, Fatou's Lemma gives

$$\int_E f \, dm \le \liminf_{n \to \infty} \int_E f_n \, dm. \tag{14.46}$$

Furthermore, we also have $f_n \le f$ a.e. on E and for all $n = 1, 2, \ldots$, so Theorem 14.3(a) implies that

$$\int_E f_n \, dm \le \int_E f \, dm$$

for all $n \in \mathbb{N}$ and then

$$\limsup_{n \to \infty} \int_E f_n \, dm \le \int_E f \, dm. \tag{14.47}$$

Hence we combine the two inequalities (14.46) and (14.47) to obtain

$$\lim_{n \to \infty} \int_E f_n \, dm = \int_E f \, dm,$$

completing the proof of the problem.   ■

Problem 14.27

$\bigstar$ $\bigstar$ *Use Fatou's Lemma to prove the Lebesgue's Dominated Convergence Theorem.*

Proof. By Problem 14.11, we observe that $f_n, f \in L^1(E)$ for every $n \in \mathbb{N}$. On the one hand, since $f_n + g \ge 0$ on E, Fatou's Lemma is applicable to gain

$$\int_E (f + g) \, dm = \int_E \liminf_{n \to \infty} (f + g) \, dm \le \liminf_{n \to \infty} \int_E (f_n + g) \, dm$$

or equivalently,

$$\int_E f \, dm \le \liminf_{n \to \infty} \int_E f_n \, dm. \tag{14.48}$$

On the other hand, since $g - f_n \ge 0$, Fatou's Lemma again shows that

$$\int_E (g - f) \, dm = \int_E \liminf_{n \to \infty} (g - f_n) \, dm \le \liminf_{n \to \infty} \int_E (g - f_n) \, dm = \int_E g \, dm - \limsup_{n \to \infty} \int_E f_n \, dm$$

so that

$$\int_E f \, dm \ge \limsup_{n \to \infty} \int_E f \, dm. \tag{14.49}$$

Hence our desired result follows by combining the inequalities (14.48) and (14.49), completing the proof of the problem.   ■

Problem 14.28

$(\star)$ Suppose that $\{f_n\}$ is a sequence of integrable functions defined on $\mathbb{R}$ and $f_n \to f$ pointwise a.e. on $\mathbb{R}$. Furthermore, for every $\epsilon > 0$, there is a measurable set $E \subseteq \mathbb{R}$ such that

$$\int_{E^c} |f_n| \, dm \le \epsilon \tag{14.50}$$

for all $n \ge N$ and

$$\int_E |f_n - f| \, dm \to 0 \tag{14.51}$$

as $n \to \infty$. Prove that

$$\lim_{n \to \infty} \int_{\mathbb{R}} |f_n - f| \, dm = 0.$$

Proof. According to Fatou's Lemma and the hypothesis (14.50), we see that

$$\int_{E^c} |f| \, dm = \int_{E^c} \liminf_{n \to \infty} |f_n| \, dm \le \liminf_{n \to \infty} \int_{E^c} |f_n| \, dm \le \epsilon. \tag{14.52}$$

Next, using the inequality (14.52), we have

$$\begin{aligned}
\int_{\mathbb{R}} |f_n - f| \, dm &\le \int_E |f_n - f| \, dm + \int_{E^c} |f_n - f| \, dm \\
&\le \int_E |f_n - f| \, dm + \int_{E^c} |f_n| \, dm + \int_{E^c} |f| \, dm \\
&\le \int_E |f_n - f| \, dm + 2\epsilon.
\end{aligned}$$

Hence we follow from the hypothesis (14.51) that

$$\limsup_{n \to \infty} \int_{\mathbb{R}} |f_n - f| \, dm \le 2\epsilon.$$

Since ϵ is arbitrary, we establish

$$\lim_{n \to \infty} \int_{\mathbb{R}} |f_n - f| \, dm = 0$$

which completes the proof of the problem. ∎

Problem 14.29

$(\star)$ Suppose that E is a measurable set, $\{f_n\}, \{g_n\}, \{h_n\} \subseteq L^1(E)$ and $g_n \le f_n \le h_n$ a.e. on E. Furthermore, we have $f_n \to f$, $g_n \to g$, $h_n \to h$ pointwise on E. If $g, h \in L^1(E)$ satisfy

$$\int_E g_n \, dm \to \int_E g \, dm \quad \text{and} \quad \int_E h_n \, dm \to \int_E h \, dm$$

as $n \to \infty$, prove that

$$\lim_{n \to \infty} \int_E f_n \, dm = \int_E f \, dm.$$

Proof. By the hypotheses, it is clear that

$$g(x) \le f(x) \le h(x)$$

a.e. on E. Since g and h are integrable on E, Problem 14.13 ensures that $f \in L^1(E)$. Now $f_n(x) - g_n(x) \ge 0$ a.e. on E for all positive integers n, so Fatou's Lemma implies that

$$\int_E f \, dm - \int_E g \, dm = \int_E (f - g) \, dm$$

$$= \int_E \liminf_{n \to \infty} (f_n - g_n) \, dm$$

$$\le \liminf_{n \to \infty} \int_E (f_n - g_n) \, dm$$

$$= \liminf_{n \to \infty} \int_E f_n \, dm - \int_E g \, dm$$

so that

$$\int_E f \, dm \le \liminf_{n \to \infty} \int_E f_n \, dm. \tag{14.53}$$

Similarly, Fatou's Lemma again shows that

$$\int_E h \, dm - \int_E f \, dm = \int_E (h - f) \, dm$$

$$= \int_E \liminf_{n \to \infty} (h_n - f_n) \, dm$$

$$\le \liminf_{n \to \infty} \int_E (h_n - f_n) \, dm$$

$$= \int_E h \, dm - \limsup_{n \to \infty} \int_E f_n \, dm$$

and then

$$\int_E f \, dm \ge \limsup_{n \to \infty} \int_E f_n \, dm. \tag{14.54}$$

Hence we derive from the inequalities (14.53) and (14.54) that

$$\lim_{n \to \infty} \int_E f_n \, dm = \int_E f \, dm.$$

This completes the proof of the problem. ∎

14.4 Applications of Convergence Theorems

Problem 14.30

(⋆) *Suppose that $f_n : E \to \mathbb{R}$ is integrable for each $n = 1, 2, \ldots$, where E is measurable with $m(E) < \infty$. Prove that if $\{f_n\}$ converges uniformly to f on E, then*

$$\lim_{n \to \infty} \int_E f_n \, dm = \int_E f \, dm. \tag{14.55}$$

Proof. Since $\{f_n\}$ converges uniformly to f on E, there exists an $N \in \mathbb{N}$ such that

$$|f_n(x) - f_N(x)| < 1$$

for all $n \geq N$ and $x \in E$. Therefore, we have

$$|f_n(x)| \leq |f_N(x)| + 1$$

on E. Since $m(E) < \infty$ and $f_N \in L^1(E)$, we conclude that $|f_N| + 1 \in L^1(E)$. Hence it follows from the Lebesgue's Dominated Convergence Theorem that the limit (14.55) holds, completing the proof of the problem.   ∎

Problem 14.31

(⋆) *Prove that*

$$\lim_{n \to \infty} \mathscr{R} \int_0^1 \frac{1 + nx^2}{(1 + x^2)^n} \, \mathrm{d}x = 0.$$

Proof. Let $f_n(x) = \frac{1+nx^2}{(1+x^2)^n}$ for every $n = 1, 2, \ldots$. By the binomial theorem, we know that

$$(1 + x^2)^n = 1 + nx^2 + \frac{n(n-1)}{2}x^4 + \cdots + x^{2n} \geq 1 + nx^2 + \frac{n(n-1)}{2}x^4$$

on $[0, 1]$ so that

$$|f_n(x)| \leq \frac{1 + nx^2}{1 + nx^2 + \frac{n(n-1)}{2}x^4} \leq 1$$

on $[0, 1]$. Therefore, each f_n is bounded on $[0, 1]$ and Theorem 14.6 ensures that $f_n \in L^1([0, 1])$ and then

$$\int_{[0,1]} f_n \, \mathrm{d}m = \mathscr{R} \int_0^1 f_n \, \mathrm{d}x.$$

Now, since we have

$$\lim_{n \to \infty} \frac{1 + nx^2}{1 + nx^2 + \frac{n(n-1)}{2}x^4} = 0$$

for every $x \in (0, 1]$, we have $f_n \to f = 0$ pointwise a.e. on $[0, 1]$. Hence we conclude from the Lebesgue's Dominated Convergence Theorem that

$$\lim_{n \to \infty} \mathscr{R} \int_0^1 \frac{1 + nx^2}{(1 + x^2)^n} \, \mathrm{d}x = \lim_{n \to \infty} \int_{[0,1]} f_n \, \mathrm{d}m = \int_{[0,1]} \lim_{n \to \infty} f_n \, \mathrm{d}m = 0$$

which ends the proof of the problem.   ∎

Problem 14.32

(⋆) *Let $-\infty < a < b < \infty$ and $f \in L^1([a, b])$. Prove that*

$$\lim_{n \to \infty} \int_{[a,b]} n \ln \left(1 + \frac{|f(x)|^2}{n^2}\right) \, \mathrm{d}m = 0.$$

Proof. Suppose that

$$f_n(x) = n \ln \left(1 + \frac{|f(x)|^2}{n^2} \right),$$

where $n = 1, 2, \ldots$. Since $f \in L^1([a,b])$, Problem 14.6 ensures that $|f|$ is finite a.e on $[a,b]$. Now it is easy to see that

$$f_n(x) = \frac{1}{n} \ln \left(1 + \frac{|f(x)|^2}{n^2} \right)^{n^2} \leq \frac{1}{n} \ln \exp(|f(x)|^2) = \frac{1}{n} |f(x)|^2 \to 0 \qquad (14.56)$$

pointwise a.e. on $[a,b]$. Next, if $y \geq 0$, then we have

$$e^{2\sqrt{y}} = 1 + 2\sqrt{y} + \frac{4y}{2} + \cdots \geq 1 + y$$

so that $\ln(1+y) \leq 2\sqrt{y}$. Thus we obtain

$$f_n(x) = n \ln \left(1 + \frac{|f(x)|^2}{n^2} \right) \leq 2n \sqrt{\frac{|f(x)|^2}{n^2}} = 2n \cdot \frac{\sqrt{|f(x)|^2}}{n} = 2|f(x)|$$

for all $n \in \mathbb{N}$ and all $x \in [a,b]$. By the fact $f \in L^1([a,b])$ and the limit (14.56), we can apply the Lebesgue's Dominated Convergence Theorem to conclude that

$$\lim_{n \to \infty} \int_{[a,b]} n \ln \left(1 + \frac{|f(x)|^2}{n^2} \right) \mathrm{d}m = \lim_{n \to \infty} \int_{[a,b]} f_n(x) \, \mathrm{d}m = \int_{[a,b]} \lim_{n \to \infty} f_n(x) \, \mathrm{d}m = 0,$$

as required. This completes the proof of the problem. $\blacksquare$

Problem 14.33

($\star$) *Suppose that $f_n, f \in L^1(\mathbb{R})$ for all $n \in \mathbb{N}$, $f_n \to f$ pointwise a.e. on $\mathbb{R}$ and*

$$\lim_{n \to \infty} \int_{\mathbb{R}} |f_n| \, \mathrm{d}m = \int_{\mathbb{R}} |f| \, \mathrm{d}m. \qquad (14.57)$$

Prove that

$$\lim_{n \to \infty} \int_E f_n \, \mathrm{d}m = \int_E f \, \mathrm{d}m$$

for every measurable subset E of $\mathbb{R}$.

Proof. Notice from the triangle inequality that

$$\big| |f_n| - |f_n - f| \big| \leq |f|,$$

so the Lebesgue's Dominated Convergence Theorem implies that

$$\lim_{n \to \infty} \left(\int_{\mathbb{R}} |f_n| \, \mathrm{d}m - \int_{\mathbb{R}} |f_n - f| \, \mathrm{d}m \right) = \lim_{n \to \infty} \int_{\mathbb{R}} (|f_n| - |f_n - f|) \, \mathrm{d}m$$

$$= \int_{\mathbb{R}} \lim_{n \to \infty} (|f_n| - |f_n - f|) \, \mathrm{d}m$$

$$= \int_{\mathbb{R}} |f| \, \mathrm{d}m. \qquad (14.58)$$

By the hypothesis (14.57) and the limit (14.58), we get

$$\int_{\mathbb{R}} |f| \, dm - \lim_{n \to \infty} \int_{\mathbb{R}} |f_n - f| \, dm = \lim_{n \to \infty} \int_{\mathbb{R}} |f_n| \, dm - \lim_{n \to \infty} \int_{\mathbb{R}} |f_n - f| \, dm$$

$$= \lim_{n \to \infty} \left(\int_{\mathbb{R}} |f_n| \, dm - \int_{\mathbb{R}} |f_n - f| \, dm \right)$$

$$= \int_{\mathbb{R}} |f| \, dm$$

which implies that

$$\lim_{n \to \infty} \int_{\mathbb{R}} |f_n - f| \, dm = 0. \tag{14.59}$$

Since $f_n - f \in L^1(\mathbb{R})$ for every positive integer n, Theorem 14.5(c) and (d) (Properties of Integrable Functions) ensures that, for every measurable subset E of $\mathbb{R}$,

$$\left| \int_E f_n \, dm - \int_E f \, dm \right| \le \int_E |f_n - f| \, dm \le \int_{\mathbb{R}} |f_n - f| \, dm. \tag{14.60}$$

Using the limit (14.59) directly to the inequality (14.60), we acquire the desired result which completes the proof of the problem. ∎

Problem 14.34

⋆ ⋆ ⋆ *Prove the Bounded Convergence Theorem.*

Proof. It is clear that the theorem holds when $m(E) = 0$. Without loss of the generality, we may assume that $m(E) > 0$ in the following discussion.

Since each f_n is measurable on E and $f_n \to f$ pointwise on E, f is measurable on E by Theorem 13.4(c) (Properties of Measurable Functions). Since $\{f_n\}$ is uniformly bounded, there exists a $M > 0$ such that

$$|f_n(x)| \le M \tag{14.61}$$

for all $n \in \mathbb{N}$ and $x \in E$. Consequently, we also have

$$|f(x)| \le M \tag{14.62}$$

for all $x \in E$. Now since $\{f_n\}$ and f are bounded on E, each $f_n - f$ is also bounded[c] on E and the paragraph following Theorem 14.5 (Properties of Integrable Functions) guarantees that

$$\left| \int_E (f_n - f) \, dm \right| \le \int_E |f_n - f| \, dm. \tag{14.63}$$

If F is any measurable subset of E, then we have

$$\int_E f_n \, dm - \int_E f \, dm = \int_E (f_n - f) \, dm$$

$$= \int_F (f_n - f) \, dm + \int_{E \setminus F} (f_n - f) \, dm,$$

---

[c]In fact, the sequence $\{f_n - f\}$ is uniformly bounded.

so we deduce from the bounds (14.61), (14.62) and the estimate (14.63) that

$$\left| \int_E f_n \, dm - \int_E f \, dm \right| \le \left| \int_F (f_n - f) \, dm \right| + \left| \int_{E \setminus F} (f_n - f) \, dm \right|$$

$$\le \int_F |f_n - f| \, dm + \int_{E \setminus F} |f_n| \, dm + \int_{E \setminus F} |f| \, dm$$

$$\le \int_F |f_n - f| \, dm + 2M \cdot m(E \setminus F). \qquad (14.64)$$

Next, given $\epsilon > 0$. Since $m(E) < \infty$, Egorov's Theorem ensures that there exists a closed set $K_\epsilon \subseteq E$ such that $m(E \setminus K_\epsilon) < \frac{\epsilon}{4M}$ and $f_n \to f$ uniformly on K_ϵ. In other words, there exists an $N \in \mathbb{N}$ such that $n \ge N$ implies

$$|f_n(x) - f(x)| < \frac{\epsilon}{2m(E)}$$

for all $x \in E$. Hence we deduce from the inequality (14.64) with F replaced by K_ϵ that

$$\left| \int_E f_n \, dm - \int_E f \, dm \right| < \frac{\epsilon}{2m(E)} \cdot m(K_\epsilon) + 2M \cdot m(E \setminus K_\epsilon) \le \frac{\epsilon}{2} + \frac{\epsilon}{2} = \epsilon$$

for all $n \ge N$. This proves the Bounded Convergence Theorem and we have completed the proof of the problem. ∎

Problem 14.35

$(\star)$ Let $-\infty < a < b < \infty$. Suppose that $f_n, f : [a, b] \to \mathbb{R}$ are measurable and $f_n(x) \ge 0$ on $[a, b]$ for every $n = 1, 2, \ldots$. If $f_n \to f$ pointwise a.e. on $[a, b]$, prove that

$$\lim_{n \to \infty} \int_{[a,b]} f_n e^{-f_n} \, dm = \int_{[a,b]} f e^{-f} \, dm.$$

Proof. It is trivial to see that $f_n e^{-f_n} \to f e^{-f}$ pointwise a.e. on $[a, b]$. Since $f_n(x) \ge 0$ on $[a, b]$, we have

$$0 \le f_n(x) e^{-f_n(x)} \le 1$$

for all $n \in \mathbb{N}$ and $x \in [a, b]$. Since $1 \in L^1([a, b])$, it yields from the Lebesgue's Dominated Convergence Theorem that

$$\lim_{n \to \infty} \int_{[a,b]} f_n e^{-f_n} \, dm = \int_{[a,b]} \lim_{n \to \infty} f_n e^{-f_n} \, dm = \int_{[a,b]} f e^{-f} \, dm,$$

completing the proof of the problem. ∎

Problem 14.36

$\star$ Let $f : \mathbb{R} \to [0, \infty)$ be measurable. If we have

$$\int_{\mathbb{R}} \frac{n^2 f(x)}{n^2 + x^2} \, dm \le 1$$

for all $n \in \mathbb{N}$, prove that

$$\int_{\mathbb{R}} f(x) \, dm \le 1.$$

Proof. Let $g_n(x) = \frac{n^2}{n^2+x^2}$, where $n \in \mathbb{N}$. It is evident that each g_n is measurable and

$$0 \le g_1(x) \le g_2(x) \le \cdots$$

for all $x \in \mathbb{R}$. Furthermore, we have $g_n(x) \to 1$ pointwise on $\mathbb{R}$. Let $f_n = g_n f$. Then $\{f_n\}$ is an increasing sequence of nonnegative measurable functions with $f_n(x) \to f(x)$ on $\mathbb{R}$, so the Lebesgue's Monotone Convergence Theorem implies that

$$\int_{\mathbb{R}} f(x) \, dm = \int_{\mathbb{R}} \lim_{n \to \infty} f_n(x) \, dm = \lim_{n \to \infty} \int_{\mathbb{R}} f_n(x) \, dm \le 1.$$

We have completed the proof of the problem. ∎

Problem 14.37

$\star$ $\star$ Prove Problem 14.20 by the Lebesgue's Monotone Convergence Theorem.

Proof. For each $n \in \mathbb{N}$, we define $A_n = \{x \in E \mid \frac{1}{n} \le |f(x)| < n\}$. Then each A_n is measurable and $A_1 \subseteq A_2 \subseteq \cdots$. Let $A = \bigcup_{n=1}^{\infty} A_n$, $A_0 = \{x \in E \mid f(x) = 0\}$ and $A_\infty = \{x \in E \mid f(x) = \infty\}$. Notice that

$$\int_E |f| \, dm = \int_{A_0} |f| \, dm + \int_{A_\infty} |f| \, dm + \int_A |f| \, dm. \tag{14.65}$$

Since $f \in L^1(E)$, Problem 14.6 ensures that $m(A_\infty) = 0$. By Theorem 14.3(f), the first two integrals on the right-hand side of the equation (14.65) are 0, thus we have

$$\int_E |f| \, dm = \int_A |f| \, dm. \tag{14.66}$$

Next, for each $n \in \mathbb{N}$, we define $f_n = |f| \chi_{A_n} : A \to \mathbb{R}$. Then $\{f_n\}$ is an increasing sequence of nonnegative measurable functions and

$$\lim_{n \to \infty} f_n(x) = |f(x)| \chi_A(x)$$

for all $x \in A$. Now the Lebesgue's Monotone Convergence Theorem implies that

$$\lim_{n \to \infty} \int_A f_n \, dm = \int_A \lim_{n \to \infty} f_n \, dm = \int_A |f| \, dm. \tag{14.67}$$

Now the two results (14.66) and (14.67) combine to give

$$\lim_{n\to\infty} \int_{A_n} |f|\,dm = \lim_{n\to\infty} \int_A f_n\,dm = \int_A |f|\,dm = \int_E |f|\,dm < \infty$$

or equivalently,

$$\lim_{n\to\infty} \int_{E\setminus A_n} |f|\,dm = 0.$$

Therefore, there is an $N \in \mathbb{N}$ such that

$$\int_{E\setminus A_N} |f|\,dm < \frac{\epsilon}{2}.$$

If we take $\delta = \frac{\epsilon}{2N}$ and F is any measurable subset of E with $m(F) < \delta$, then Theorem 14.5(c) and (d) (Properties of Integrable Functions) say that

$$\left| \int_F f\,dm \right| \le \int_F |f|\,dm$$
$$\le \int_{(E\setminus A_N)\cap F} |f|\,dm + \int_{A_N\cap F} |f|\,dm$$
$$\le \int_{E\setminus A_N} |f|\,dm + \int_{A_N\cap F} N\,dm$$
$$< \frac{\epsilon}{2} + N \cdot m(A_N \cap F)$$
$$< \frac{\epsilon}{2} + N \cdot \frac{\epsilon}{2N}$$
$$= \epsilon.$$

This completes the proof of the problem.   ■

Problem 14.38

(⋆) *Does the Lebesgue's Monotone Convergence Theorem hold for decreasing sequences of measurable functions?*

Proof. The answer is negative. In fact, for each $n \in \mathbb{N}$, we consider $f_n = \chi_{[n,\infty)}$ defined on $\mathbb{R}$. Then $\{f_n\}$ is clearly a decreasing sequence of nonnegative measurable functions and

$$f_n(x) \to f = 0$$

pointwise on $\mathbb{R}$. However, we note that

$$\int_{\mathbb{R}} f_n\,dm = \int_n^\infty dx = \infty \ne 0 = \int_{\mathbb{R}} f\,dm,$$

so the Lebesgue's Monotone Convergence Theorem fails in this case. We complete the proof of the problem.   ■

Problem 14.39

$(\star)(\star)$ *Suppose that $f \in L^1(\mathbb{R})$. Prove that*

$$\lim_{n \to \infty} \int_{\mathbb{R}} \frac{xf(x-n)}{1+|x|}\, dm = \int_{\mathbb{R}} f\, dm.$$

Proof. Let n be a fixed positive integer. By the change of variable (x replaced by $x - n$), we see immediately that

$$\int_{\mathbb{R}} \frac{xf(x-n)}{1+|x|}\, dm = \int_{-n}^{\infty} \frac{(x+n)f(x)}{1+(x+n)}\, dm + \int_{-\infty}^{-n} \frac{(x+n)f(x)}{1-(x+n)}\, dm. \qquad (14.68)$$

Thus we are going to compute the two integrals on the right-hand side of the equation (14.68) when $n \to \infty$.

To compute the first integral, we notice that if $x \geq -n$, then we have

$$0 \leq \frac{x+n}{1+(x+n)} < 1 \quad \text{and} \quad \lim_{n \to \infty} \frac{x+n}{1+(x+n)} = 1.$$

Therefore, we get

$$0 \leq \left| \frac{(x+n)f(x)}{1+(x+n)} \right| \leq |f(x)| \qquad (14.69)$$

on $[-n, \infty)$. Define $g_n : \mathbb{R} \to \mathbb{R}$ by

$$g_n(x) = \frac{(x+n)f(x)}{1+(x+n)} \chi_{[-n,\infty)}(x),$$

so the inequalities (14.69) show that

$$|g_n| \leq |f| \quad \text{and} \quad g_n \to f \text{ pointwise a.e. on } \mathbb{R}.$$

Since $|f| \in L^1(\mathbb{R})$ by Theorem 14.5(d) (Properties of Integrable Functions), it follows from the Lebesgue's Dominated Convergence Theorem that

$$\lim_{n \to \infty} \int_{-n}^{\infty} \frac{(x+n)f(x)}{1+(x+n)}\, dm = \lim_{n \to \infty} \int_{\mathbb{R}} g_n\, dm = \int_{\mathbb{R}} \lim_{n \to \infty} g_n\, dm = \int_{\mathbb{R}} f\, dm. \qquad (14.70)$$

Next, we define $h_n : \mathbb{R} \to \mathbb{R}$ by

$$h_n(x) = \frac{(x+n)f(x)}{1-(x+n)} \chi_{(-\infty,-n]}(x).$$

Now if $x \leq -n$, then we have

$$0 \leq \left| \frac{(x+n)}{1-(x+n)} \right| < 1 \quad \text{and} \quad \lim_{n \to \infty} \frac{(x+n)}{1-(x+n)} = -1.$$

Therefore, they yield that

$$|h_n| \leq |f| \quad \text{and} \quad h_n \to 0 \text{ pointwise a.e. on } \mathbb{R}.$$

Again the Lebesgue's Dominated Convergence Theorem guarantees that

$$\lim_{n\to\infty} \int_{-\infty}^{-n} \frac{(x+n)f(x)}{1-(x+n)}\,dm = \lim_{n\to\infty} \int_{\mathbb{R}} h_n\,dm = \int_{\mathbb{R}} \lim_{n\to\infty} h_n\,dm = 0. \qquad (14.71)$$

Finally, we substitute the two limits (14.70) and (14.71) back into the equation (14.68) to conclude that

$$\lim_{n\to\infty} \int_{\mathbb{R}} \frac{xf(x-n)}{1+|x|}\,dm = \lim_{n\to\infty} \int_{-n}^{\infty} \frac{(x+n)f(x)}{1+(x+n)}\,dm + \lim_{n\to\infty} \int_{-\infty}^{-n} \frac{(x+n)f(x)}{1-(x+n)}\,dm = \int_{\mathbb{R}} f\,dm.$$

This completes the proof of the problem.   ∎

Problem 14.40

(⋆) Let $E \subseteq \mathbb{R}$ and $f : E \to \mathbb{R}$ be measurable and $E_n = \{x \in E \,|\, |f(x)| \geq n\}$, where n is a positive integer. If $f \in L^1(E)$. prove that

$$\lim_{n\to\infty} \int_{E_n} |f|\,dm = 0.$$

Proof. Since E is measurable, every E_n is also measurable. Let $f_n = |f|\chi_{E_n} : E \to \mathbb{R}$ for each $n = 1, 2, \ldots$. Clearly, each f_n is measurable. Since $f \in L^1(E)$, Problem 14.6 ensures that f is finite a.e. on E and then we have $|f_n(x)| \leq |f(x)|$ for all $n \in \mathbb{N}$ and all $x \in E$. In addition, it is easy to check that

$$f_n(x) \to 0$$

pointwise a.e. on E. By applying the Lebesgue's Dominated Convergence Theorem, we conclude that

$$\lim_{n\to\infty} \int_{E_n} |f|\,dm = \lim_{n\to\infty} \int_{E} f_n\,dm = \int_{E} \lim_{n\to\infty} f_n\,dm = 0.$$

Therefore, it completes the proof of the problem.   ∎

Problem 14.41

(⋆) Suppose that $f \in L^1(\mathbb{R})$. Prove that the series

$$\sum_{n=-\infty}^{\infty} f(x+n)$$

converges absolutely a.e. on $\mathbb{R}$.

Proof. We define

$$F(x) = \sum_{n=-\infty}^{\infty} |f(x+n)| \geq 0.$$

We claim that F is finite a.e. on $\mathbb{R}$. In fact, since F is periodic of period 1, it suffices to show that F is finite a.e. on $[0,1]$. Applying the Lebesgue's Monotone Convergence Theorem for Series and then Theorem 14.3(e), we get

$$
\int_{[0,1]} F(x)\,\mathrm{d}m = \int_{[0,1]} \Big(\sum_{n=-\infty}^{\infty} |f(x+n)| \Big)\,\mathrm{d}m
$$

$$
= \sum_{n=-\infty}^{\infty} \int_{[0,1]} |f(x+n)|\,\mathrm{d}m
$$

$$
= \sum_{n=-\infty}^{\infty} \int_{[n,n+1]} |f(x)|\,\mathrm{d}m
$$

$$
= \int_{\mathbb{R}} |f|\,\mathrm{d}m
$$

$$
< \infty.
$$

In other words, F is integrable on $[0,1]$ and it deduces from Problem 14.6 that F is finite a.e. on $[0,1]$, as claimed. This completes the proof of the problem. ∎

Problem 14.42

⭐ ⭐ Suppose that $E \subseteq \mathbb{R}$ is measurable and $f \in L^1(E)$. Given $\epsilon > 0$. Prove that there exists a measurable set F with finite measure such that

$$
\sup_{x \in F} |f(x)| < \infty \quad \text{and} \quad \int_{F^c} |f|\,\mathrm{d}m < \epsilon.
$$

Proof. For each $n \in \mathbb{N}$, let $F_n = \{x \in E \mid \frac{1}{n} \le |f(x)| \le n\} \subseteq E$. Now we deduce from Problem 14.5 that

$$
m(F_n) \le m\Big(\Big\{x \in E \,\Big|\, |f(x)| \ge \frac{1}{n}\Big\}\Big) \le n \int_{F_n} |f|\,\mathrm{d}m \le n \int_E |f|\,\mathrm{d}m
$$

which reduces to

$$
\frac{m(F_n)}{n} \le \int_E |f|\,\mathrm{d}m < \infty
$$

for every $n = 1, 2, \ldots$. Thus for each *fixed* positive integer n, if we set $F = F_n$, then we have

$$
m(F) < \infty \quad \text{and} \quad \sup_{x \in F} |f(x)| \le n < \infty.
$$

To find the estimate of the integral, we let $f_n = f \chi_{F_n^c} : E \to \mathbb{R}$ for each $n = 1, 2, \ldots$. Then we have $|f_n| \le |f|$ on E. By the definition, we see that

$$
F_n^c = \Big\{x \in E \,\Big|\, 0 \le |f(x)| < \frac{1}{n} \text{ or } |f(x)| > n\Big\}.
$$

Recall from Problem 14.6 that f is finite a.e. on E, so we conclude that $f_n(x) \to 0$ pointwise a.e. on E. Consequently, the Lebesgue's Dominated Convergence Theorem is applicable and we get

$$
\lim_{n \to \infty} \int_E f_n\,\mathrm{d}m = \int_E \lim_{n \to \infty} f_n\,\mathrm{d}m = 0.
$$

Therefore, there exists an $N \in \mathbb{N}$ such that

$$\int_E f_N \, dm < \epsilon. \tag{14.72}$$

Finally, we know from the definition of f_N that the integral (14.72) is equal to

$$\int_{F_N^c} |f| \, dm.$$

We have completed the proof of the problem.   ∎

CHAPTER 15

Differential Calculus of Functions of Several Variables

▌15.1 Fundamental Concepts

The definition and basic properties of derivatives of $f : \mathbb{R} \to \mathbb{R}$ have been introduced and discussed in [25, Chap. 8]. This chapter extends the differential calculus theory to functions from $\mathbb{R}^n$ to $\mathbb{R}^m$ and the main references for this chapter are [2, Chap. 12], [18, Chap. 9], [22, Chap. 6] and [27, Chap. 7 & 8]. Here we assume that you are familiar with elementary linear algebra which can be found, for examples, in [10] or [13].

▌15.1.1 The Directional Derivatives and the Partial Derivatives

Let S be a subset of $\mathbb{R}^n$, $\mathbf{x}$ be an interior point of S, $\mathbf{v}$ be a point of S and $\mathbf{f} : S \to \mathbb{R}^m$ be a function.

Definition 15.1 (Directional Derivatives). *The **directional derivative** of **f** at **x** in the direction **v**, denoted by $\mathbf{f}'_{\mathbf{v}}(\mathbf{x})$ or $\nabla_{\mathbf{v}}\mathbf{f}(\mathbf{x})$, is defined by the limit*

$$\mathbf{f}'_{\mathbf{v}}(\mathbf{x}) = \lim_{h \to 0} \frac{\mathbf{f}(\mathbf{x} + h\mathbf{v}) - \mathbf{f}(\mathbf{x})}{h}$$

whenever the limit exists.[a]

Let $\{\mathbf{e}_1, \mathbf{e}_2, \ldots, \mathbf{e}_n\}$ and $\{\mathbf{u}_1, \mathbf{u}_2, \ldots, \mathbf{u}_m\}$ be the standard bases of $\mathbb{R}^n$ and $\mathbb{R}^m$ respectively, x_j be the jth component of $\mathbf{x}$ with respect to the standard basis $\{\mathbf{e}_1, \mathbf{e}_2, \ldots, \mathbf{e}_n\}$ and

$$\mathbf{f} = (f_1, f_2, \ldots, f_m).$$

Then we have

$$\mathbf{f}(\mathbf{x}) = \sum_{i=1}^{m} f_i(\mathbf{x})\mathbf{u}_i \quad \text{and} \quad f_i(\mathbf{x}) = \mathbf{f}(\mathbf{x}) \cdot \mathbf{u}_i.$$

---

[a]Since $\mathbf{x} \in S^\circ$, it is true that $\mathbf{x} + h\mathbf{v} \in S$ for all small $h > 0$ so that $\mathbf{f}'_{\mathbf{v}}(\mathbf{x})$ is well-defined.

Besides, it is clear from Definition 15.1 (Directional Derivatives) that $\mathbf{f}'_{\mathbf{v}}(\mathbf{x})$ exists if and only if all $(f_i)'_{\mathbf{v}}(\mathbf{x})$ exist, where $i = 1, 2, \ldots, m$. In this case, we have

$$\mathbf{f}'_{\mathbf{v}}(\mathbf{x}) = ((f_1)'_{\mathbf{v}}(\mathbf{x}), (f_2)'_{\mathbf{v}}(\mathbf{x}), \ldots, (f_m)'_{\mathbf{v}}(\mathbf{x})).$$

Definition 15.2 (Partial Derivatives). *For $1 \leq i \leq m$ and $1 \leq j \leq n$, we define*

$$(D_j f_i)(\mathbf{x}) = \lim_{h \to 0} \frac{f_i(\mathbf{x} + h\mathbf{e}_j) - f_i(\mathbf{x})}{h}$$

*whenever the limit exists. Then each $(D_j f_i)(\mathbf{x})$ is called the **partial derivative** of f_i with respect to x_j.*[b]

Remark 15.1

It is very easy to check from the above two definitions that the existence of $\mathbf{f}'_{\mathbf{v}}(\mathbf{x})$ for all $\mathbf{v} \in \mathbb{R}^n$ implies the existence of $(D_j f_i)(\mathbf{x})$, where $1 \leq i \leq m$ and $1 \leq j \leq n$. Particularly, if $\mathbf{v} = \mathbf{e}_j$, then we have

$$\mathbf{f}'_{\mathbf{e}_j}(\mathbf{x}) = ((D_j f_1)(\mathbf{x}), (D_j f_2)(\mathbf{x}), \ldots, (D_j f_m)(\mathbf{x})),$$

where $1 \leq j \leq n$. However, the converse is false, see Problem 15.2 below.

15.1.2 Differentiation of Functions of Several Variables

Definition 15.3. *The function $\mathbf{f} : S \subseteq \mathbb{R}^n \to \mathbb{R}^m$ is said to be **differentiable at** $\mathbf{x} \in \mathbb{R}^n$ if there exists a **linear transformation** $T_{\mathbf{x}} : \mathbb{R}^n \to \mathbb{R}^m$ such that*

$$\lim_{\mathbf{h} \to 0} \frac{|\mathbf{f}(\mathbf{x} + \mathbf{h}) - \mathbf{f}(\mathbf{x}) - T_{\mathbf{x}}(\mathbf{h})|}{|\mathbf{h}|} = 0. \tag{15.1}$$

*In this case, we write $\mathbf{f}'(\mathbf{x}) = T_{\mathbf{x}}$. If $\mathbf{f}$ is differentiable at every point of S, then we say that $\mathbf{f}$ is **differentiable in** S.*[c]

The derivative $\mathbf{f}'(\mathbf{x})$ (or equivalently the transformation $T_{\mathbf{x}}$) is called the **differential of f at x** or the **total derivative of f at x**. Furthermore, the limit (15.1) is equivalent to the form

$$\mathbf{f}(\mathbf{x} + \mathbf{h}) - \mathbf{f}(\mathbf{x}) = \mathbf{f}'(\mathbf{x})\mathbf{h} + \mathbf{R}(\mathbf{h}), \tag{15.2}$$

where the remainder $\mathbf{R}(\mathbf{h})$ satisfies

$$\lim_{\mathbf{h} \to 0} \frac{|\mathbf{R}(\mathbf{h})|}{|\mathbf{h}|} = 0. \tag{15.3}$$

Here we have some basic properties of the total derivative $\mathbf{f}'(\mathbf{x})$.

Theorem 15.4 (Uniqueness of the Total Derivative). *Suppose that both the linear transformations $T_{\mathbf{x}}$ and $\widehat{T}_{\mathbf{x}}$ satisfy Definition 15.3. Then we have $T_{\mathbf{x}} = \widehat{T}_{\mathbf{x}}$.*

---

[b]Another notation for $D_j f_i$ is $\dfrac{\partial f_i}{\partial x_j}$.

[c]Reader should pay attention that the $\mathbf{h}$ is a vector, *not* a scalar.

Theorem 15.5. *If* **f** *is differentiable at* $\mathbf{p} \in S \subseteq \mathbb{R}^n$, *then* **f** *is continuous at* **p**.

> **Remark 15.2**
>
> A consequence of Theorem 15.5 is that directional differentiability *does not* imply total differentiability, see Problem 15.3. However, they are connected in *some* way. In fact, Problem 15.4 ensures that the existence of $\mathbf{f}'(\mathbf{x})$ implies that of $\mathbf{f}'_\mathbf{v}(\mathbf{x})$ for every $\mathbf{v} \in \mathbb{R}^n$.

Theorem 15.6 (The Chain Rule). *Suppose that* S *is open in* $\mathbb{R}^n$, $\mathbf{f} : S \to \mathbb{R}^m$ *is differentiable at* $\mathbf{a}$ *and* $\mathbf{g} : \mathbf{f}(S) \subseteq \mathbb{R}^m \to \mathbb{R}^k$ *is differentiable at* $\mathbf{f}(\mathbf{a})$. *Then the mapping* $\mathbf{F} = \mathbf{g} \circ \mathbf{f} : S \to \mathbb{R}^k$ *is differentiable at* $\mathbf{a}$ *and*

$$\mathbf{F}'(\mathbf{a}) = \mathbf{g}'(\mathbf{f}(\mathbf{a})) \times \mathbf{f}'(\mathbf{a}). \tag{15.4}$$

We note that on the right-hand side of the equation (15.4), we have **the product of two linear transformations**.

15.1.3 The Total Derivatives and the Jacobian Matrices

Theorem 15.7. *Let* $\{\mathbf{e}_1, \mathbf{e}_2, \ldots, \mathbf{e}_n\}$ *and* $\{\mathbf{u}_1, \mathbf{u}_2, \ldots, \mathbf{u}_m\}$ *denote the standard bases of* $\mathbb{R}^n$ *and* $\mathbb{R}^m$ *respectively. Suppose that* S *is open in* $\mathbb{R}^n$ *and* $\mathbf{f} : S \subseteq \mathbb{R}^n \to \mathbb{R}^m$ *is differentiable at* $\mathbf{x}$. *Then all the partial derivatives* $(D_j f_i)(\mathbf{x})$ *exist, where* $1 \le i \le m$ *and* $1 \le j \le n$, *and furthermore the total derivative* $\mathbf{f}'(\mathbf{x})$ *satisfies*

$$\mathbf{f}'(\mathbf{x})\mathbf{e}_j = \sum_{i=1}^m (D_j f_i)(\mathbf{x})\mathbf{u}_i, \tag{15.5}$$

where $1 \le j \le n$.

This theorem says that, once we know that $\mathbf{f}'(\mathbf{x})$ exists, we can express it in terms of the partial derivatives $(D_j f_i)(\mathbf{x})$. In addition, we recall from Definition 15.3 that $\mathbf{f}'(\mathbf{x})$ is in fact a linear transformation $T_\mathbf{x} : \mathbb{R}^n \to \mathbb{R}^m$, so elementary linear algebra tells us that there is a (unique) matrix associated with $\mathbf{f}'(\mathbf{x})$. This matrix is called the **Jacobian matrix** of **f** at **x** and it is denoted by $\mathbf{J_f}(\mathbf{x})$. We see from the forms (15.5) that $\mathbf{J_f}(\mathbf{x})$ is related to the partial derivatives $(D_j f_i)(\mathbf{x})$ as follows:

$$\mathbf{J_f}(\mathbf{x}) = \begin{pmatrix} (D_1 f_1)(\mathbf{x}) & (D_2 f_1)(\mathbf{x}) & \cdots & (D_n f_1)(\mathbf{x}) \\ (D_1 f_2)(\mathbf{x}) & (D_2 f_2)(\mathbf{x}) & \cdots & (D_n f_2)(\mathbf{x}) \\ \vdots & \vdots & \ddots & \vdots \\ (D_1 f_m)(\mathbf{x}) & (D_2 f_m)(\mathbf{x}) & \cdots & (D_n f_m)(\mathbf{x}) \end{pmatrix}_{m \times n}. \tag{15.6}$$

In terms of the Jacobian matrices, the matrix form of Theorem 15.6 (The Chain Rule) can be written as

$$\mathbf{J_F}(\mathbf{x}) = \mathbf{J_g}(\mathbf{f}(\mathbf{x})) \times \mathbf{J_f}(\mathbf{x}), \tag{15.7}$$

where this time the product on the right-hand side of the equation (15.7) refers to matrix product.

In addition, if $m = 1$, then $\mathbf{f} = f$ is real-valued so the formula (15.5) gives

$$f'(\mathbf{x})\mathbf{v} = v_1 f'(\mathbf{x})\mathbf{e}_1 + v_2 f'(\mathbf{x})\mathbf{e}_2 + \ldots + v_n f'(\mathbf{x})\mathbf{e}_n$$

$$= v_1(D_1f)(\mathbf{x}) + v_2(D_2f)(\mathbf{x}) + \cdots + v_n(D_nf)(\mathbf{x})$$
$$= \nabla f(\mathbf{x}) \cdot \mathbf{v},$$

where

$$\nabla f(\mathbf{x}) = ((D_1f)(\mathbf{x}), (D_2f)(\mathbf{x}), \dots, (D_nf)(\mathbf{x})) \tag{15.8}$$

is called the **gradient of f at x**.

15.1.4   The Mean Value Theorem for Differentiable Functions

The Mean Value Theorem for Differentiable Functions. *Suppose that S is open in $\mathbb{R}^n$ and $\mathbf{f} : S \to \mathbb{R}^m$ is differentiable in S. If $\lambda \mathbf{x} + (1 - \lambda)\mathbf{y} \in S$, where $0 \le \lambda \le 1$, then for every $\mathbf{a} \in \mathbb{R}^m$, there exists a point z lying on the line segment of $\mathbf{x}$ and $\mathbf{y}$ such that*

$$\mathbf{a} \cdot [\mathbf{f}(\mathbf{y}) - \mathbf{f}(\mathbf{x})] = \mathbf{a} \cdot [\mathbf{f}'(\mathrm{z})(\mathbf{y} - \mathbf{x})].$$

Theorem 15.8. *Suppose that the open set S in the previous theorem is convex and there is a constant M such that*

$$\sup_{|\mathbf{y}| \le 1} \{|\mathbf{f}'(\mathbf{x})\mathbf{y}|\} \le M$$

for every $\mathbf{x} \in S$.[d] *Then we have*

$$|\mathbf{f}(\mathbf{b}) - \mathbf{f}(\mathbf{a})| \le M|\mathbf{b} - \mathbf{a}|$$

for every $\mathbf{a}, \mathbf{b} \in S$.

Theorem 15.9. *Let $S \subseteq \mathbb{R}^n$ be open and convex. If $\mathbf{f} : S \to \mathbb{R}^m$ is differentiable in S and $\mathbf{f}'(\mathbf{x}) = \mathbf{0}$ for all $\mathbf{x} \in S$, then $\mathbf{f}$ is a constant on S.*

15.1.5   Continuously Differentiable Functions

Definition 15.10 (Continuously Differentiable Functions). *Denote $L(\mathbb{R}^n, \mathbb{R}^m)$ to be the set of all linear transformations of $\mathbb{R}^n$ into $\mathbb{R}^m$. Let $S \subseteq \mathbb{R}^n$ be open and $\mathbf{f} : S \to \mathbb{R}^m$ be differentiable. We say that $\mathbf{f}$ is **continuously differentiable** in S if*

$$\mathbf{f}' : S \to L(\mathbb{R}^n, \mathbb{R}^m)$$

is a continuous mapping. The class of all continuously differentiable functions in S is denoted by $\mathscr{C}'(S)$.

Now the following theorem tells us a necessary and sufficient condition for $\mathbf{f} \in \mathscr{C}'(S)$.

Theorem 15.11. *Suppose that S is open in $\mathbb{R}^n$. Then $\mathbf{f} : S \to \mathbb{R}^m$ belongs to $\mathscr{C}'(S)$ if and only if all D_jf_i exist and are continuous on S for all $1 \le i \le m$ and $1 \le j \le n$.*

---

[d]Note that $\mathbf{f}'(\mathbf{x})\mathbf{y}$ is a vector in $\mathbb{R}^m$.

15.1.6 The Inverse Function Theorem and the Implicit Function Theorem

The Inverse Function Theorem. *Let E be an open subset of $\mathbb{R}^n$ and $\mathbf{f} : E \to \mathbb{R}^n$ be a **continuously differentiable** function on E. Let $\mathbf{a} \in E$ be a point such that the linear transformation $\mathbf{f}'(\mathbf{a})$ is invertible, i.e., $\det \mathbf{J_f}(\mathbf{a}) \neq 0$. Then there exists an open set $U \subseteq E$ containing $\mathbf{a}$ and an open set $V \subseteq \mathbb{R}^n$ containing $\mathbf{b} = \mathbf{f}(\mathbf{a})$ such that $\mathbf{f} : U \to V$ is a bijection. Furthermore, the inverse mapping $\mathbf{f}^{-1} : V \to U$ is differentiable at $\mathbf{x}$ and*

$$(\mathbf{f}^{-1})'(\mathbf{y}) = \frac{1}{\mathbf{f}'(\mathbf{x})}$$

for every $\mathbf{x} \in U$ and $\mathbf{y} = \mathbf{f}(\mathbf{x})$.

See Figure 15.1 for the idea of this theorem.

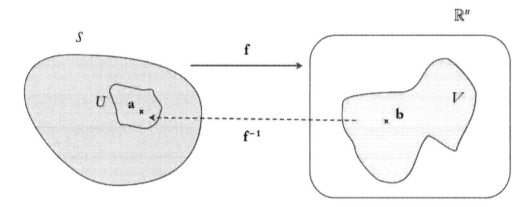

Figure 15.1: The Inverse Function Theorem.

The Implicit Function Theorem. *Let E be an open subset of $\mathbb{R}^{n+m}$ and $\mathbf{f} : E \to \mathbb{R}^m$ be a **continuously differentiable** function on E. We write $(\mathbf{x}, \mathbf{y}) = (x_1, x_2, \dots, x_n, y_1, y_2, \dots, y_m)$. Let $(\mathbf{a}, \mathbf{b}) \in E$ be a point such that $\mathbf{f}(\mathbf{a}, \mathbf{b}) = \mathbf{0}$, where $\mathbf{a} \in \mathbb{R}^n$ and $\mathbf{b} \in \mathbb{R}^m$. Suppose that*

$$\det \begin{pmatrix} \dfrac{\partial f_1}{\partial y_1}(\mathbf{a},\mathbf{b}) & \dfrac{\partial f_1}{\partial y_2}(\mathbf{a},\mathbf{b}) & \cdots & \dfrac{\partial f_1}{\partial y_m}(\mathbf{a},\mathbf{b}) \\[2mm] \dfrac{\partial f_2}{\partial y_1}(\mathbf{a},\mathbf{b}) & \dfrac{\partial f_2}{\partial y_2}(\mathbf{a},\mathbf{b}) & \cdots & \dfrac{\partial f_2}{\partial y_m}(\mathbf{a},\mathbf{b}) \\[2mm] \vdots & \vdots & \ddots & \vdots \\[2mm] \dfrac{\partial f_m}{\partial y_1}(\mathbf{a},\mathbf{b}) & \dfrac{\partial f_m}{\partial y_2}(\mathbf{a},\mathbf{b}) & \cdots & \dfrac{\partial f_m}{\partial y_m}(\mathbf{a},\mathbf{b}) \end{pmatrix}_{m \times m} \neq 0.$$

Then there exists an open set $U \subseteq \mathbb{R}^n$ containing $\mathbf{a}$ such that there exists a unique continuously differentiable function $\mathbf{g} : U \to \mathbb{R}^m$ satisfying

$$\mathbf{g}(\mathbf{a}) = \mathbf{b} \quad and \quad \mathbf{f}(\mathbf{x}, \mathbf{g}(\mathbf{x})) = \mathbf{0}$$

for all $\mathbf{x} \in U$. *Furthermore, we have*

$$
\mathbf{g}'(\mathbf{x}) = - \begin{pmatrix} \dfrac{\partial f_1}{\partial y_1}(\mathbf{x}, \mathbf{g}(\mathbf{x})) & \cdots & \dfrac{\partial f_1}{\partial y_m}(\mathbf{x}, \mathbf{g}(\mathbf{x})) \\ \vdots & \ddots & \vdots \\ \dfrac{\partial f_m}{\partial y_1}(\mathbf{x}, \mathbf{g}(\mathbf{x})) & \cdots & \dfrac{\partial f_m}{\partial y_m}(\mathbf{x}, \mathbf{g}(\mathbf{x})) \end{pmatrix}_{m \times m}^{-1}
$$

$$
\times \begin{pmatrix} \dfrac{\partial f_1}{\partial x_1}(\mathbf{x}, \mathbf{g}(\mathbf{x})) & \cdots & \dfrac{\partial f_1}{\partial x_n}(\mathbf{x}, \mathbf{g}(\mathbf{x})) \\ \vdots & \ddots & \vdots \\ \dfrac{\partial f_m}{\partial x_1}(\mathbf{x}, \mathbf{g}(\mathbf{x})) & \cdots & \dfrac{\partial f_m}{\partial x_n}(\mathbf{x}, \mathbf{g}(\mathbf{x})) \end{pmatrix}_{m \times n}
$$

for all $\mathbf{x} \in U$.

Remark 15.3

Classically, mathematicians prove first the Inverse Function Theorem and then derive from it the Implicit Function Theorem. In addition, the usual ingredients of proofs of the two theorems consist of compactness of balls in $\mathbb{R}^n$ and the Contraction Principle (see [18, Theorem 9.23, p. 220] or [22, Theorem 6.6.4, p. 150]). In 2013, Oliveira [14] employed Dini's approach to prove first the Implicit Function Theorem by induction and then he derived from it the Inverse Function Theorem. The tools he used are the Intermediate Value Theorem and the Mean Value Theorem for Derivatives. For an extensive treatment of the Implicit Function Theorem, we refer the reader to the book [8].

▌15.1.7 Higher Order Derivatives

The partial derivatives $D_1\mathbf{f}, D_2\mathbf{f}, \ldots, D_n\mathbf{f}$ of a function $\mathbf{f} : \mathbb{R}^n \to \mathbb{R}^m$ are also functions from $\mathbb{R}^n$ to $\mathbb{R}^m$. If the partial derivatives $D_j\mathbf{f}$ are differentiable, then they are called the **second-order partial derivatives of f** which are denoted by

$$
D_{ij}\mathbf{f} = D_i(D_j\mathbf{f}) = \frac{\partial}{\partial x_i}\left(\frac{\partial \mathbf{f}}{\partial x_j}\right) = \frac{\partial^2 \mathbf{f}}{\partial x_i \partial x_j},
$$

where $1 \le i, j \le n$.

Theorem 15.12 (Clairaut's Theorem)**.** *If both partial derivatives $D_i\mathbf{f}$ and $D_j\mathbf{f}$ exist in an open set $S \subseteq \mathbb{R}^n$ and both $D_{ij}\mathbf{f}$ and $D_{ji}\mathbf{f}$ are continuous at $\mathbf{p} \in S$, then we have*

$$
(D_{ij}\mathbf{f})(\mathbf{p}) = (D_{ji}\mathbf{f})(\mathbf{p}).
$$

15.2 Differentiation of Functions of Several Variables

Problem 15.1

$(\star)$ *Consider the function $f : \mathbb{R}^2 \to \mathbb{R}$ defined by*

$$f(x,y) = \begin{cases} 1, & \text{if } xy = 0; \\ \\ 0, & \text{otherwise.} \end{cases}$$

Show that $(D_1 f)(0,0) = (D_2 f)(0,0)$, but f is discontinuous at $(0,0)$.

Proof. We note that

$$(D_1 f)(0,0) = \lim_{h \to 0} \frac{f(h,0) - f(0,0)}{h} = \lim_{h \to 0} \frac{1-1}{h} = 0$$

and

$$(D_2 f)(0,0) = \lim_{h \to 0} \frac{f(0,h) - f(0,0)}{h} = \lim_{h \to 0} \frac{1-1}{h} = 0.$$

Thus $(D_1 f)(0,0) = (D_2 f)(0,0)$. However, we notice that

$$\lim_{x \to 0} f(x,x) = 0 \neq 1 = f(0,0)$$

which means that f is discontinuous at $(0,0)$. This completes the proof of the problem. ∎

Remark 15.4

Problem 15.1 asserts that the existence of the partial derivatives $D_1 f, D_2 f, \ldots, D_n f$ at $\mathbf{p}$ *does not* necessarily imply the continuity of f at $\mathbf{p}$. Fortunately, this really happens when all partial derivatives are bounded in an open set $E \subseteq \mathbb{R}^n$, see [18, Exercise 7, p. 239].

Problem 15.2

$(\star)$ *Consider $f : \mathbb{R}^2 \to \mathbb{R}$ which is given by*

$$f(x,y) = \begin{cases} -x - y, & \text{if } xy = 0; \\ \\ 1, & \text{otherwise.} \end{cases}$$

Prove that both $(D_1 f)(\mathbf{0})$ and $(D_2 f)(\mathbf{0})$ exist, but not $f_{\mathbf{v}}(\mathbf{0})$ for any $\mathbf{v} = (a,b)$ with $ab \neq 0$.

Proof. By Definition 15.2 (Partial Derivatives), we have

$$(D_1 f)(\mathbf{0}) = \lim_{h \to 0} \frac{f(h,0) - f(0,0)}{h} = \lim_{h \to 0} \frac{-h}{h} = -1$$

and

$$(D_2 f)(\mathbf{0}) = \lim_{h \to 0} \frac{f(0,h) - f(0,0)}{h} = \lim_{h \to 0} \frac{-h}{h} = -1.$$

However, if $\mathbf{v} = (a,b)$, where $ab \neq 0$, then $a \neq 0$ and $b \neq 0$. Therefore, we obtain from Definition 15.1 (Directional Derivatives) that

$$f'_{\mathbf{v}}(\mathbf{0}) = \lim_{h \to 0} \frac{f(ha, hb) - f(0,0)}{h} = \lim_{h \to 0} \frac{1}{h}$$

which does not exist. This completes the proof of the problem.   ■

Problem 15.3

(⋆) *Prove that the function $f : \mathbb{R}^2 \to \mathbb{R}$ defined by*

$$f(x,y) = \begin{cases} \dfrac{xy^2}{x^2 + y^6}, & \text{if } x \neq 0; \\ \\ 0, & \text{otherwise.} \end{cases}$$

has well-defined $f'_{\mathbf{v}}(\mathbf{0})$ for every $\mathbf{v}$ but f is discontinuous at $\mathbf{0}$.

Proof. Suppose that $\mathbf{v} = (a,b)$. By Definition 15.1 (Directional Derivatives), we have

$$\begin{aligned} f'_{\mathbf{v}}(\mathbf{0}) &= \lim_{h \to 0} \frac{f(\mathbf{0} + h\mathbf{v}) - f(\mathbf{0})}{h} \\ &= \lim_{h \to 0} \frac{f(ha, hb)}{h} \\ &= \lim_{h \to 0} \frac{ha \cdot h^2 b^2}{h(h^2 a^2 + h^6 b^6)} \\ &= \lim_{h \to 0} \frac{ab^2}{a^2 + h^4 b^6} \\ &= \begin{cases} \dfrac{b^2}{a}, & \text{if } a \neq 0; \\ \\ 0, & \text{otherwise.} \end{cases} \end{aligned}$$

Thus $f'_{\mathbf{v}}(\mathbf{0})$ is well-defined for every $\mathbf{v}$.

To show that f is discontinuous at $\mathbf{0}$, we first note that $f(0,0) = 0$. Next, if $y \neq 0$, then we have

$$f(y^2, y) = \frac{y^4}{y^4 + y^6} = \frac{1}{1 + y^2}$$

which implies that

$$\lim_{y \to 0} f(y^2, y) = 1 \neq f(0,0).$$

Hence f is discontinuous at $\mathbf{0}$ and we complete the proof of the problem.   ■

Problem 15.4

($\star$) *Suppose that* $\mathbf{x}$ *is an interior point of* $S \subseteq \mathbb{R}^n$, $\mathbf{f} : S \subseteq \mathbb{R}^n \to \mathbb{R}^m$ *is a function and* $\mathbf{v} \in \mathbb{R}^n$. *If* $\mathbf{f}$ *is differentiable at* $\mathbf{x}$, *prove that*

$$\mathbf{f}'_{\mathbf{v}}(\mathbf{x}) = \mathbf{f}'(\mathbf{x})\mathbf{v}. \tag{15.9}$$

Proof. Fix $\mathbf{v}$ and consider $\mathbf{h} = t\mathbf{v}$ for some sufficiently small real t. Since $\mathbf{f}$ is differentiable at $\mathbf{x}$, it follows from the approximation (15.2) that

$$\begin{aligned}
\mathbf{f}(\mathbf{x} + t\mathbf{v}) - \mathbf{f}(\mathbf{x}) - t\mathbf{f}'(\mathbf{x})\mathbf{v} &= \mathbf{f}(\mathbf{x} + t\mathbf{v}) - \mathbf{f}(\mathbf{x}) - \mathbf{f}'(\mathbf{x})(t\mathbf{v}) \\
&= \mathbf{f}(\mathbf{x} + \mathbf{h}) - \mathbf{f}(\mathbf{x}) - \mathbf{f}'(\mathbf{x})\mathbf{h} \\
&= \mathbf{R}(\mathbf{h}) \\
&= \mathbf{R}(t\mathbf{v}).
\end{aligned}$$

By the limit (15.3), we see that

$$\lim_{t \to 0} \frac{|\mathbf{R}(t\mathbf{v})|}{|t|} = 0.$$

Thus we get

$$\mathbf{f}'_{\mathbf{v}}(\mathbf{x}) - \mathbf{f}'(\mathbf{x})\mathbf{v} = \lim_{t \to 0} \frac{\mathbf{f}(\mathbf{x} + t\mathbf{v}) - \mathbf{f}(\mathbf{x}) - t\mathbf{f}'(\mathbf{x})\mathbf{v}}{t} = \lim_{t \to 0} \frac{\mathbf{R}(t\mathbf{v})}{t} = \mathbf{0},$$

i.e., $\mathbf{f}'_{\mathbf{v}}(\mathbf{x}) = \mathbf{f}'(\mathbf{x})\mathbf{v}$. This completes the proof of the problem. ∎

Remark 15.5

We notice that the left-hand side of the formula (15.9) is a limit, while its right-hand side is a matrix product. In other words, this gives an easier way (matrix product) to compute the directional derivative $\mathbf{f}'_{\mathbf{v}}(\mathbf{x})$ whenever $\mathbf{f}$ is differentiable at $\mathbf{x}$.

Problem 15.5

($\star$) *Suppose that* $\mathbf{f}'_{\mathbf{v}}(\mathbf{x})$ *and* $\mathbf{g}'_{\mathbf{v}}(\mathbf{x})$ *exist. Prove that* $(\mathbf{f} + \mathbf{g})'_{\mathbf{v}}(\mathbf{x}) = \mathbf{f}'_{\mathbf{v}}(\mathbf{x}) + \mathbf{g}'_{\mathbf{v}}(\mathbf{x})$.

Proof. By Definition 15.1 (Directional Derivatives), we have

$$\begin{aligned}
(\mathbf{f} + \mathbf{g})'_{\mathbf{v}}(\mathbf{x}) &= \lim_{h \to 0} \frac{(\mathbf{f} + \mathbf{g})(\mathbf{x} + h\mathbf{v}) - (\mathbf{f} + \mathbf{g})(\mathbf{x})}{h} \\
&= \lim_{h \to 0} \frac{\mathbf{f}(\mathbf{x} + h\mathbf{v}) - \mathbf{f}(\mathbf{x})}{h} + \lim_{h \to 0} \frac{\mathbf{g}(\mathbf{x} + h\mathbf{v}) - \mathbf{g}(\mathbf{x})}{h} \\
&= \mathbf{f}'_{\mathbf{v}}(\mathbf{x}) + \mathbf{g}'_{\mathbf{v}}(\mathbf{x}),
\end{aligned}$$

completing the proof of the problem. ∎

Problem 15.6

⋆ Prove Theorem 15.5.

Proof. Since $\mathbf{f}'(\mathbf{p}) = T_{\mathbf{p}}$, we rewrite the formula (15.2) as

$$\mathbf{f}(\mathbf{p}+\mathbf{h}) - \mathbf{f}(\mathbf{p}) = T_{\mathbf{p}}(\mathbf{h}) + \mathbf{R}(\mathbf{h}). \tag{15.10}$$

Let $\mathbf{h} = h_1\mathbf{e}_1 + h_2\mathbf{e}_2 + \cdots + h_n\mathbf{e}_n$. Then $\mathbf{h} \to \mathbf{0}$ if and only if $h_i \to 0$ for all $i = 1, 2, \ldots, n$. Since $T_{\mathbf{p}}$ is linear, we have

$$T_{\mathbf{p}}(\mathbf{h}) = h_1 T_{\mathbf{p}}(\mathbf{e}_1) + h_2 T_{\mathbf{p}}(\mathbf{e}_2) + \cdots + h_n T_{\mathbf{p}}(\mathbf{e}_n)$$

so that $T_{\mathbf{p}}(\mathbf{h}) \to \mathbf{0}$ as $\mathbf{h} \to \mathbf{0}$. By the limit (15.3), we see that $\mathbf{R}(\mathbf{h}) \to \mathbf{0}$ as $\mathbf{h} \to \mathbf{0}$. Hence we conclude from the formula (15.10) that

$$\lim_{\mathbf{h} \to \mathbf{0}} \mathbf{f}(\mathbf{p}+\mathbf{h}) = \mathbf{f}(\mathbf{p}),$$

i.e., $\mathbf{f}$ is continuous at $\mathbf{p}$. This completes the proof of the problem. ∎

Problem 15.7

⋆ Prove Theorem 15.7.

Proof. By Problem 15.4 and then Remark 15.1, we obtain

$$\mathbf{f}'(\mathbf{x})\mathbf{e}_j = \mathbf{f}'_{\mathbf{e}_j}(\mathbf{x}) = ((D_j f_1)(\mathbf{x}), (D_j f_2)(\mathbf{x}), \ldots, (D_j f_m)(\mathbf{x})) = \sum_{i=1}^{m}(D_j f_i)(\mathbf{x})\mathbf{u}_i.$$

We have completed the proof of the problem. ∎

Problem 15.8

⋆ ⋆ Let $S \subseteq \mathbb{R}^n$ and $\mathbf{x}$ be an interior point of S. Suppose that $\mathbf{f} : S \to \mathbb{R}^m$ is given by $\mathbf{f} = (f_1, f_2, \ldots, f_m)$. Use Definition 15.3 to prove that $\mathbf{f}$ is differentiable at $\mathbf{x}$ if and only if every $f_i : S \to \mathbb{R}$ is differentiable at $\mathbf{x}$.

Proof. We notice that for any $\mathbf{y} = (y_1, y_2, \ldots, y_m) \in \mathbb{R}^m$, we have

$$|y_i| \leq |\mathbf{y}| \leq \sqrt{m} \max_{1 \leq i \leq m} |y_i| \tag{15.11}$$

for all $i = 1, 2, \ldots, m$. Now the Jacobian matrices for the component functions f_i, where $i = 1, 2, \ldots, m$, at $\mathbf{x}$ are given by

$$\mathbf{J}_{f_i}(\mathbf{x}) = \begin{pmatrix} (D_1 f_i)(\mathbf{x}) & (D_2 f_i)(\mathbf{x}) & \cdots & (D_n f_i)(\mathbf{x}) \end{pmatrix}.$$

If we compare these with the matrix (15.6), then we assert that

$$\mathbf{J_f}(\mathbf{x}) = \begin{pmatrix} \mathbf{J}_{f_1}(\mathbf{x}) \\ \mathbf{J}_{f_2}(\mathbf{x}) \\ \vdots \\ \mathbf{J}_{f_m}(\mathbf{x}) \end{pmatrix} \quad \text{or} \quad \mathbf{f}'(\mathbf{x}) = \begin{pmatrix} f_1'(\mathbf{x}) \\ f_2'(\mathbf{x}) \\ \vdots \\ f_m'(\mathbf{x}) \end{pmatrix}.$$

Therefore, the ith component function of $\mathbf{f}(\mathbf{x} + \mathbf{h}) - \mathbf{f}(\mathbf{x}) - \mathbf{f}'(\mathbf{x})\mathbf{h}$ is exactly

$$f_i(\mathbf{x} + \mathbf{h}) - f_i(\mathbf{x}) - f_i'(\mathbf{x})\mathbf{h}$$

and then we follow from the inequalities (15.11) that

$$\frac{|f_i(\mathbf{x} + \mathbf{h}) - f_i(\mathbf{x}) - f_i'(\mathbf{x})\mathbf{h}|}{|\mathbf{h}|} \leq \frac{|\mathbf{f}(\mathbf{x} + \mathbf{h}) - \mathbf{f}(\mathbf{x}) - \mathbf{f}'(\mathbf{x})\mathbf{h}|}{|\mathbf{h}|}$$
$$\leq \sqrt{m} \times \frac{|f_i(\mathbf{x} + \mathbf{h}) - f_i(\mathbf{x}) - f_i'(\mathbf{x})\mathbf{h}|}{|\mathbf{h}|}. \tag{15.12}$$

Hence, by taking $\mathbf{h} \to \mathbf{0}$ to the inequalities (15.12), we get the desired result. This completes the proof of the problem. ∎

Problem 15.9

(⋆) *Let* $T : \mathbb{R}^n \to \mathbb{R}^m$ *be linear. Prove that* T *is differentiable everywhere on* $\mathbb{R}^n$ *and* $T'(\mathbf{x}) = T$.

Proof. Since T is linear, we have

$$T(\mathbf{x} + \mathbf{h}) - T(\mathbf{x}) - T(\mathbf{h}) = \mathbf{0}$$

which is in the form (15.2). Next Theorem 15.4 (Uniqueness of the Total Derivative) implies that

$$T'(\mathbf{x}) = T$$

which ends the proof of the problem. ∎

Problem 15.10

(⋆) *Use Theorem 15.6 (The Chain Rule) and Problem 15.9 to prove Problem 15.8 again.*

Proof. Suppose that $\mathbf{f}$ is differentiable at $\mathbf{x}$. Let $\pi_i : \mathbb{R}^m \to \mathbb{R}$ be the projection

$$\pi_i(\mathbf{y}) = \pi_i(y_1, y_2, \ldots, y_m) = y_i,$$

where $1 \leq i \leq m$. Since each π_i is linear, every π_i is differentiable everywhere on $\mathbb{R}^m$ by Problem 15.9. Obviously, we have $f_i = \pi_i \circ \mathbf{f}$, so Theorem 15.6 (The Chain Rule) implies that f_i is also differentiable at $\mathbf{x}$.

Conversely, we suppose that every f_i is differentiable at $\mathbf{x} \in S$. Define $\phi_i : \mathbb{R} \to \mathbb{R}^m$ by

$$\phi_i(y) = (0, \ldots, 0, y, 0, \ldots, 0),$$

where the y is the ith coordinate. It is obvious that each ϕ_i is linear so that it is differentiable everywhere on $\mathbb{R}$ by Problem 15.9. By Theorem 15.6 (The Chain Rule), the composite function $\phi_i \circ f_i : S \to \mathbb{R}^m$ is differentiable at $\mathbf{x}$. Since

$$\mathbf{f} = (f_1, f_2, \ldots, f_m) = \sum_{i=1}^{m} \phi_i \circ f_i,$$

we have the desired result that $\mathbf{f}$ is differentiable at $\mathbf{x}$. This completes the proof of the problem. ∎

Problem 15.11

⭐ *Suppose that $\mathbf{f}, \mathbf{g} : \mathbb{R}^n \to \mathbb{R}^m$ are functions such that $\mathbf{f}$ is differentiable at $\mathbf{p}$, $\mathbf{f}(\mathbf{p}) = \mathbf{0}$ and $\mathbf{g}$ is continuous at $\mathbf{p}$. If $\mathbf{h}(\mathbf{x}) = \mathbf{g}(\mathbf{x}) \cdot \mathbf{f}(\mathbf{x})$, prove that*

$$\mathbf{h}'(\mathbf{p})\mathbf{v} = \mathbf{g}(\mathbf{p}) \cdot [\mathbf{f}'(\mathbf{p})\mathbf{v}], \tag{15.13}$$

where $\mathbf{v} \in \mathbb{R}^n$.

Proof. Since $\mathbf{f}$ is differentiable at $\mathbf{p}$ and $\mathbf{f}(\mathbf{p}) = \mathbf{0}$, we get from the formula (15.2) that

$$\mathbf{f}(\mathbf{p} + \mathbf{h}) = \mathbf{f}'(\mathbf{p})\mathbf{h} + \mathbf{R}(\mathbf{h})$$

which gives

$$
\begin{aligned}
\mathbf{h}(\mathbf{p} + \mathbf{h}) - \mathbf{h}(\mathbf{p}) &= \mathbf{g}(\mathbf{p} + \mathbf{h}) \cdot \mathbf{f}(\mathbf{p} + \mathbf{h}) - \mathbf{g}(\mathbf{p}) \cdot \mathbf{f}(\mathbf{p}) \\
&= \mathbf{g}(\mathbf{p} + \mathbf{h}) \cdot [\mathbf{f}'(\mathbf{p})\mathbf{h} + \mathbf{R}(\mathbf{h})] \\
&= \mathbf{g}(\mathbf{p} + \mathbf{h}) \cdot [\mathbf{f}'(\mathbf{p})\mathbf{h} + \mathbf{R}(\mathbf{h})] + \mathbf{g}(\mathbf{p}) \cdot \mathbf{f}'(\mathbf{p})\mathbf{h} - \mathbf{g}(\mathbf{p}) \cdot \mathbf{f}'(\mathbf{p})\mathbf{h} \\
&= \mathbf{g}(\mathbf{p}) \cdot \mathbf{f}'(\mathbf{p})\mathbf{h} + \mathbf{g}(\mathbf{p} + \mathbf{h}) \cdot \mathbf{R}(\mathbf{h}) + [\mathbf{g}(\mathbf{p} + \mathbf{h}) - \mathbf{g}(\mathbf{p})] \cdot \mathbf{f}'(\mathbf{p})\mathbf{h}. \tag{15.14}
\end{aligned}
$$

Since $\mathbf{g}$ is continuous at $\mathbf{p}$, the limit (15.3) implies

$$\lim_{\mathbf{h} \to 0} \frac{|\mathbf{g}(\mathbf{p} + \mathbf{h}) \cdot \mathbf{R}(\mathbf{h})|}{|\mathbf{h}|} = |\mathbf{g}(\mathbf{p})| \lim_{\mathbf{h} \to 0} \frac{|\mathbf{R}(\mathbf{h})|}{|\mathbf{h}|} = 0.$$

Furthermore, we see that

$$
\begin{aligned}
\lim_{\mathbf{h} \to 0} \frac{[\mathbf{g}(\mathbf{p} + \mathbf{h}) - \mathbf{g}(\mathbf{p})] \cdot \mathbf{f}'(\mathbf{p})\mathbf{h}}{|\mathbf{h}|} &= \lim_{\mathbf{h} \to 0} [\mathbf{g}(\mathbf{p} + \mathbf{h}) - \mathbf{g}(\mathbf{p})] \cdot \mathbf{f}'(\mathbf{p})\left(\frac{\mathbf{h}}{|\mathbf{h}|}\right) \\
&= \mathbf{0} \cdot \lim_{\mathbf{h} \to 0} \mathbf{f}'(\mathbf{p})\left(\frac{\mathbf{h}}{|\mathbf{h}|}\right) \\
&= 0.
\end{aligned}
$$

Thus the equation (15.14) can be expressed as

$$\mathbf{h}(\mathbf{p} + \mathbf{h}) - \mathbf{h}(\mathbf{p}) = \mathbf{g}(\mathbf{p}) \cdot \mathbf{f}'(\mathbf{p})\mathbf{h} + \widehat{\mathbf{R}}(\mathbf{h}),$$

where

$$\lim_{h \to 0} \frac{|\widehat{\mathbf{R}}(\mathbf{h})|}{|\mathbf{h}|} = 0.$$

By Definition 15.3, we have the expected result (15.13), completing the proof of the problem. ∎

Problem 15.12

$(\star)$ *Suppose that* $\mathbf{f} : \mathbb{R}^3 \to \mathbb{R}^2$ *is given by*

$$\mathbf{f}(x, y, z) = (x^2 - y + z, e^{-xy} \sin z + xz).$$

Prove that $\mathbf{f}$ *is differentiable at* $\mathbf{x} \in \mathbb{R}^3$ *and find* $\mathbf{J_f}(\mathbf{p})$.

Proof. We notice that the component functions f_1 and f_2 of $\mathbf{f}$ are given by

$$f_1(x, y, z) = x^2 - y + z \quad \text{and} \quad f_2(x, y, z) = e^{-xy} \sin z + xz$$

so that

$$
\begin{aligned}
D_1 f_1 &= 2x, \quad D_2 f_1 = -1, \quad D_3 f_1 = 1, \\
D_1 f_2 &= -y e^{-xy} \sin z + z, \quad D_2 f_2 = -x e^{-xy} \sin z, \quad D_3 f_2 = e^{-xy} \cos z + x.
\end{aligned}
\tag{15.15}
$$

Since all $D_j f_i$ exist and are continuous on $\mathbb{R}^3$, we deduce from Theorem 15.11 that $\mathbf{f}$ is differentiable in $\mathbb{R}^3$. Furthermore, we gain immediately from the partial derivatives (15.15) that

$$\mathbf{J_f}(\mathbf{x}) = \begin{pmatrix} 2x & -1 & 1 \\ -y e^{-xy} \sin z + z & -x e^{-xy} \sin z & e^{-xy} \cos z + x \end{pmatrix}.$$

This completes the analysis of the problem. ∎

Problem 15.13

$(\star)$ *Suppose that* $f, g, h : \mathbb{R}^2 \to \mathbb{R}$ *are differentiable such that* $z = f(x, y), x = g(r, \theta)$ *and* $y = h(r, \theta)$. *Verify that*

$$\frac{\partial z}{\partial r} = \frac{\partial z}{\partial x} \cdot \frac{\partial x}{\partial r} + \frac{\partial z}{\partial y} \cdot \frac{\partial y}{\partial r} \quad \text{and} \quad \frac{\partial z}{\partial \theta} = \frac{\partial z}{\partial x} \cdot \frac{\partial x}{\partial \theta} + \frac{\partial z}{\partial y} \cdot \frac{\partial y}{\partial \theta}. \tag{15.16}$$

Proof. Let $z = \psi(r, \theta) = f(g(r, \theta), h(r, \theta))$ and $\phi = (g, h)$. Since $z = f(x, y)$, the matrix form (15.6) or the gradient form (15.8) shows that

$$\mathbf{J}_f = \begin{pmatrix} \dfrac{\partial f}{\partial x} & \dfrac{\partial f}{\partial y} \end{pmatrix} = \begin{pmatrix} \dfrac{\partial z}{\partial x} & \dfrac{\partial z}{\partial y} \end{pmatrix}.$$

Similarly, we have

$$\mathbf{J}_\psi = \begin{pmatrix} \dfrac{\partial z}{\partial r} & \dfrac{\partial z}{\partial \theta} \end{pmatrix} \quad \text{and} \quad \mathbf{J}_\phi = \begin{pmatrix} \dfrac{\partial x}{\partial r} & \dfrac{\partial x}{\partial \theta} \\ \dfrac{\partial y}{\partial r} & \dfrac{\partial y}{\partial \theta} \end{pmatrix}.$$

Since $\psi = f \circ \phi$, the matrix form (15.7) gives $\mathbf{J}_\psi = \mathbf{J}_f \times \mathbf{J}_\phi$ or explicitly,

$$
\begin{pmatrix} \dfrac{\partial z}{\partial r} & \dfrac{\partial z}{\partial \theta} \end{pmatrix} = \begin{pmatrix} \dfrac{\partial z}{\partial x} & \dfrac{\partial z}{\partial y} \end{pmatrix} \times \begin{pmatrix} \dfrac{\partial x}{\partial r} & \dfrac{\partial x}{\partial \theta} \\[2mm] \dfrac{\partial y}{\partial r} & \dfrac{\partial y}{\partial \theta} \end{pmatrix}
$$

$$
= \begin{pmatrix} \dfrac{\partial z}{\partial x} \cdot \dfrac{\partial x}{\partial r} + \dfrac{\partial z}{\partial y} \cdot \dfrac{\partial y}{\partial r} & \dfrac{\partial z}{\partial x} \cdot \dfrac{\partial x}{\partial \theta} + \dfrac{\partial z}{\partial y} \cdot \dfrac{\partial y}{\partial \theta} \end{pmatrix}
$$

which is exactly the formulas (15.16), completing the analysis of the problem. ∎

Problem 15.14

⋆ Let $\mathbf{p} \in S \subseteq \mathbb{R}^n$. *Prove that there is no function* $f : S \to \mathbb{R}$ *such that* $f'_{\mathbf{v}}(\mathbf{p}) > 0$ *for every* $\mathbf{v} \in \mathbb{R}^n \setminus \{\mathbf{0}\}$.

Proof. Assume that there was a function $f : S \to \mathbb{R}$ such that $f'_{\mathbf{v}}(\mathbf{p}) > 0$ for every $\mathbf{v} \in \mathbb{R}^n \setminus \{\mathbf{0}\}$. Let $\alpha \in \mathbb{R} \setminus \{0\}$. Then we notice from Definition 15.1 (Directional Derivatives) that

$$
f'_{\alpha \mathbf{v}}(\mathbf{p}) = \lim_{h \to 0} \frac{f(\mathbf{p} + h\alpha\mathbf{v}) - f(\mathbf{p})}{h} = \alpha \lim_{\alpha h \to 0} \frac{f(\mathbf{p} + h\alpha\mathbf{v}) - f(\mathbf{p})}{\alpha h} = \alpha f'_{\mathbf{v}}(\mathbf{p}).
$$

In particular, we have

$$
f'_{-\mathbf{v}}(\mathbf{p}) = -f'_{\mathbf{v}}(\mathbf{p}) < 0
$$

for every $\mathbf{v} \in \mathbb{R}^n \setminus \{\mathbf{0}\}$, a contradiction. Hence we end the proof of the problem. ∎

Problem 15.15

⋆ *Suppose that* $f, g, h, k : S \subseteq \mathbb{R}^n \to \mathbb{R}$ *are functions such that* $h = fg$ *and* $k = \frac{f}{g}$. *Find* ∇h *and* ∇k.

Proof. By the definition (15.8), we have

$$
\nabla h = (D_1 h, D_2 h, \dots, D_n h). \tag{15.17}
$$

For $1 \le i \le n$, since $D_i h = D_i(fg) = (D_i f)g + f(D_i g)$, we obtain from the expression (15.17) that

$$
\begin{aligned}
\nabla h &= ((D_1 f)g + f(D_1 g), (D_2 f)g + f(D_2 g), \dots, (D_n f)g + f(D_n g)) \\
&= (D_1 f, D_2 f, \dots, D_n f)g + f(D_1 g, D_2 g, \dots, D_n g) \\
&= (\nabla f)g + f(\nabla g). \tag{15.18}
\end{aligned}
$$

Write $k = fg^{-1}$, so the formula (15.18) gives

$$
\nabla k = (\nabla f)g^{-1} + f[\nabla(g^{-1})]. \tag{15.19}
$$

By the definition (15.8) again, we know that

$$
\nabla(g^{-1}) = (D_1(g^{-1}), D_2(g^{-1}), \dots, D_n(g^{-1})) = \left(\frac{-D_1 g}{g^2}, \frac{-D_2 g}{g^2}, \dots, \frac{-D_n g}{g^2} \right) = -\frac{\nabla g}{g^2}. \tag{15.20}
$$

If we substitute the expression (15.20) into the formula (15.19), then we assert that

$$\nabla k = \frac{\nabla f}{g} - \frac{f \nabla g}{g^2} = \frac{1}{g^2} (g \nabla f - f \nabla g).$$

This completes the analysis of the problem.   ■

Problem 15.16

($\star$) *Suppose that $f : \mathbb{R} \to \mathbb{R}$ is differentiable in $\mathbb{R}$ and $g : \mathbb{R}^3 \to \mathbb{R}$ is defined by*

$$g(x, y, z) = x^k + y^k + z^k,$$

where $k \in \mathbb{N}$. Denote $h = f \circ g$. Prove that

$$|\nabla h(x, y, z)|^2 = k^2 [x^{2(k-1)} + y^{2(k-1)} + z^{2(k-1)}] \times [f'(g(x, y, z))]^2. \qquad (15.21)$$

Proof. We have $h(x, y, z) = f(x^k + y^k + z^k)$, so Theorem 15.6 (The Chain Rule) implies

$$D_1 h = k x^{k-1} f'(x^k + y^k + z^k).$$

Similarly, we have

$$D_2 h = k y^{k-1} f'(x^k + y^k + z^k) \quad \text{and} \quad D_3 h = k z^{k-1} f'(x^k + y^k + z^k).$$

Thus it follows from the formula (15.8) that

$$\nabla h(x, y, z) = (k x^{k-1} f'(x^k + y^k + z^k), k y^{k-1} f'(x^k + y^k + z^k), k z^{k-1} f'(x^k + y^k + z^k))$$

and then

$$\begin{aligned} |\nabla h(x, y, z)|^2 &= k^2 [x^{2(k-1)} + y^{2(k-1)} + z^{2(k-1)}] \times [f'(x^k + y^k + z^k)]^2 \\ &= k^2 [x^{2(k-1)} + y^{2(k-1)} + z^{2(k-1)}] \times [f'(g(x, y, z)]^2 \end{aligned}$$

which is the desired result (15.21). This ends the proof of the problem.   ■

15.3 The Mean Value Theorem for Differentiable Functions

Problem 15.17

($\star$) ($\star$) *Prove the Mean Value Theorem for Differentiable Functions.*

Proof. Fix $\mathbf{a} \in \mathbb{R}^m$. Define the function $F : [0, 1] \to \mathbb{R}$ by

$$F(t) = \mathbf{a} \cdot \mathbf{f}(\mathbf{x} + t(\mathbf{y} - \mathbf{x})). \qquad (15.22)$$

The hypothesis shows that

$$\mathbf{x} + t(\mathbf{y} - \mathbf{x}) = (1 - t)\mathbf{x} + t\mathbf{y} \in S$$

for all $t \in [0, 1]$ so that the function F is well-defined. Since $\mathbf{f}$ is differentiable in S, Theorem 15.5 guarantees that $\mathbf{f}$ is continuous on S and then F must be continuous on $[0, 1]$. Furthermore, we apply Theorem 15.6 (The Chain Rule) to the definition (15.22) to get

$$F'(t) = \mathbf{a} \cdot \mathbf{f}'(\mathbf{x} + t(\mathbf{y} - \mathbf{x}))(\mathbf{y} - \mathbf{x}).$$

Now the Mean Value Theorem for Derivatives [25, p. 129] implies the existence of a $\xi \in (0, 1)$ such that

$$F(1) - F(0) = F'(\xi) = \mathbf{a} \cdot [\mathbf{f}'(\mathbf{x} + \xi(\mathbf{y} - \mathbf{x}))(\mathbf{y} - \mathbf{x})]. \tag{15.23}$$

We know that $F(1) = \mathbf{a} \cdot \mathbf{f}(\mathbf{y})$ and $F(0) = \mathbf{a} \cdot \mathbf{f}(\mathbf{x})$, so if we let $\mathbf{z} = \mathbf{x} + \xi(\mathbf{y} - \mathbf{x}) \in S$, then the formula (15.23) can be rewritten as

$$\mathbf{a} \cdot [\mathbf{f}(\mathbf{y}) - \mathbf{f}(\mathbf{x})] = \mathbf{a} \cdot [\mathbf{f}'(\mathbf{z})(\mathbf{y} - \mathbf{x})].$$

This completes the proof of the problem.   ■

Problem 15.18

$\circledast$ Let S be open and convex in $\mathbb{R}^n$ and $f : S \to \mathbb{R}$. Let f be differentiable in S and $\mathbf{a}, \mathbf{a} + \mathbf{h} \in S$. Prove that there exists a $\lambda \in (0, 1)$ such that

$$f(\mathbf{a} + \mathbf{h}) - f(\mathbf{a}) = f'(\mathbf{a} + \lambda \mathbf{h})\mathbf{h}.$$

Proof. Let $c \neq 0$. Since S is convex, we have $\mathbf{a} + \lambda \mathbf{h} = \lambda(\mathbf{a} + \mathbf{h}) + (1 - \lambda)\mathbf{a} \in S$, where $0 \le \lambda \le 1$. By the Mean Value Theorem for Differentiable Functions, we get

$$c[f(\mathbf{a} + \mathbf{h}) - f(\mathbf{a})] = c[f'(\mathbf{z})(\mathbf{a}\mathbf{h} - \mathbf{a})] = c[f'(\mathbf{z})\mathbf{h}] \tag{15.24}$$

for some $\mathbf{z} = \mathbf{a} + \lambda \mathbf{h}$. Hence our result follows from dividing the expression (15.24) by the nonzero constant c and we have completed the proof of the problem.   ■

Problem 15.19

$\circledast$ Prove Theorem 15.9.

Proof. Since $\mathbf{f}'(\mathbf{x}) = \mathbf{0}$ on S, we have $|\mathbf{f}'(\mathbf{x})\mathbf{y}| = 0$ for every $|\mathbf{y}| \le 1$. Thus we may take $M = 0$ in Theorem 15.8 to conclude that

$$|\mathbf{f}(\mathbf{b}) - \mathbf{f}(\mathbf{a})| = 0$$

for all $\mathbf{a}, \mathbf{b} \in S$. This means that $\mathbf{f}$ is constant on S, completing the proof of the problem.   ■

Problem 15.20

$\circledast$ $\circledast$ Let $\mathbf{a} \in S \subseteq \mathbb{R}^n$ and $B_r(\mathbf{a}) = \{\mathbf{x} \in S \,|\, |\mathbf{x} - \mathbf{a}| < r\}$, where $r > 0$. Suppose that $f : S \to \mathbb{R}$ satisfies $f'_{\mathbf{v}}(\mathbf{x}) = 0$ for every $\mathbf{x} \in B_r(\mathbf{a})$ and every $\mathbf{v} \in \mathbb{R}^n$. Prove that f is constant on $B_r(\mathbf{a})$.

Proof. We first claim that if the directional derivative $f_{\mathbf{v}}'(\mathbf{a} + t\mathbf{v})$ exists for each $t \in [0, 1]$, then there exists a $\lambda \in (0, 1)$ such that

$$f(\mathbf{a} + \mathbf{v}) - f(\mathbf{a}) = f_{\mathbf{v}}'(\mathbf{a} + \lambda\mathbf{v}). \tag{15.25}$$

To this end, we define $g : [0, 1] \to \mathbb{R}$ by

$$g(t) = f(\mathbf{a} + t\mathbf{v}). \tag{15.26}$$

By Definition 15.1 (Directional Derivatives), we see that

$$g'(t) = \lim_{h \to 0} \frac{g(t + h) - g(t)}{h} = \lim_{h \to 0} \frac{f(\mathbf{a} + t\mathbf{v} + h\mathbf{v}) - f(\mathbf{a} + t\mathbf{v})}{h} = f_{\mathbf{v}}'(\mathbf{a} + t\mathbf{v}), \tag{15.27}$$

where $t \in [0, 1]$. Therefore, the Mean Value Theorem for Derivatives shows that there exists a $\lambda \in (0, 1)$ such that

$$g(1) - g(0) = g'(\lambda)(1 - 0) = g'(\lambda). \tag{15.28}$$

By the definition (15.26), we have $g(1) = f(\mathbf{a} + \mathbf{v})$ and $g(0) = f(\mathbf{a})$. Hence our desired result (15.25) follows directly from the comparison of the expressions (15.27) and (15.28).

Next, for every $\mathbf{x} \in B_r(\mathbf{a})$, let $\mathbf{v} = \mathbf{x} - \mathbf{a}$ so that $|\mathbf{v}| < r$. Since $|\mathbf{a} + \lambda\mathbf{v} - \mathbf{a}| = \lambda|\mathbf{v}| < \lambda r < r$, the hypothesis shows that $f_{\mathbf{v}}'(\mathbf{a} + \lambda\mathbf{v}) = 0$ and then

$$f(\mathbf{x}) = f(\mathbf{a} + \mathbf{v}) = f(\mathbf{a}).$$

In other words, f is constant on $B_r(\mathbf{a})$ which completes the proof of the problem. ∎

15.4 The Inverse Function Theorem and the Implicit Function Theorem

---
Problem 15.21

(⋆) Suppose that $E = \{(x, y) \in \mathbb{R}^2 \,|\, x > y\}$. Define $\mathbf{f} : E \to \mathbb{R}^2$ by

$$\mathbf{f}(x, y) = (f_1(x, y), f_2(x, y)) = (x + y, x^2 + y^2).$$

Prove that $\mathbf{f}$ is locally bijective.

Proof. Since $D_1 f_1 = 1$, $D_2 f_1 = 1$, $D_1 f_2 = 2x$ and $D_2 f_2 = 2y$ are all continuous on E, Theorem 15.11 implies that $\mathbf{f} \in \mathscr{C}'(E)$. Furthermore, we have

$$\mathbf{J}_{\mathbf{f}}(x, y) = \begin{pmatrix} 1 & 1 \\ 2x & 2y \end{pmatrix}.$$

Since $x > y$ if $(x, y) \in E$, it is true that $\det \mathbf{J}_{\mathbf{f}}(x, y) = 2(y - x) \neq 0$. Hence it deduces from the Inverse Function Theorem that $\mathbf{f}$ is locally bijective. This completes the analysis of the problem. ∎

Problem 15.22

(⋆) Suppose that $\mathbf{f} : B_r(\mathbf{a}) \to \mathbb{R}^n$ is differentiable at $\mathbf{a}$ and its inverse $\mathbf{f}^{-1}$ exists and is differentiable at $\mathbf{f}(\mathbf{a})$. Prove that $\det \mathbf{J_f}(\mathbf{a}) \neq 0$.

Proof. Assume that $\det \mathbf{J_f}(\mathbf{a}) = 0$. By Problem 15.9 and Theorem 15.6 (The Chain Rule), we can establish

$$I_n = (\mathbf{f}^{-1} \circ \mathbf{f})(\mathbf{a}) = (\mathbf{f}^{-1} \circ \mathbf{f})'(\mathbf{a}) = (\mathbf{f}^{-1})'(\mathbf{f}(\mathbf{a})) \times \mathbf{f}'(\mathbf{a})$$

so that

$$1 = \det I_n = \det \mathbf{J_{f^{-1}}}(\mathbf{f}(\mathbf{a})) \times \det \mathbf{J_f}(\mathbf{a}) = 0,$$

a contradiction. Hence we must have $\det \mathbf{J_f}(\mathbf{a}) \neq 0$ and we complete the proof of the problem. ∎

Remark 15.6

In other words, Problem 15.22 says that the hypothesis $\det \mathbf{J_f}(\mathbf{a}) \neq 0$ cannot be omitted in the Inverse Function Theorem.

Problem 15.23

(⋆) Show, by a counterexample, that the hypothesis $\mathbf{f} \in \mathscr{C}'(E)$ cannot be dropped in the Inverse Function Theorem.

Proof. We consider the function $f : \mathbb{R} \to \mathbb{R}$ defined by

$$f(x) = \begin{cases} x + 2x^2 \sin \dfrac{1}{x}, & \text{if } x \neq 0; \\ \\ 0, & \text{otherwise.} \end{cases}$$

Obviously, f is differentiable in $(-1, 1)$ and

$$f'(0) = \lim_{h \to 0} \frac{f(h) - f(0)}{h} = \lim_{h \to 0} \left(1 + 2h \sin \frac{1}{h} \right) = 1 \neq 0.$$

However, for $x \neq 0$, we have

$$f'(x) = 1 + 4x \sin \frac{1}{x} - 2 \cos \frac{1}{x}$$

so that $\lim_{x \to 0} f'(x)$ does not exist. In other words, $f \notin \mathscr{C}'((-1, 1))$.

Let V be any neighborhood of 0. Then there exists an $N \in \mathbb{N}$ such that $\frac{2}{(4N-3)\pi} \in V$. It is clear that V also contains $\frac{2}{(4N-1)\pi}$ and $\frac{2}{(4N+1)\pi}$ because

$$\frac{2}{(4N+1)\pi} < \frac{2}{(4N-1)\pi} < \frac{2}{(4N-3)\pi}.$$

By direct computation, we see that

$$f\left(\frac{2}{(4N-1)\pi} \right) < f\left(\frac{2}{(4N+1)\pi} \right) < f\left(\frac{2}{(4N-3)\pi} \right). \tag{15.29}$$

Now the inequalities (15.29) and the continuity of f imply that f is *not* injective in V. Thus the hypothesis $\mathbf{f} \in \mathscr{C}'(E)$ cannot be dropped in the Inverse Function Theorem and it completes the proof of the problem. ∎

Problem 15.24

$(\star)$ $(\star)$ *Suppose that* $\mathbf{f} : \mathbb{R}^m \to \mathbb{R}^m$ *is an element of* $\mathscr{C}'(\mathbb{R}^m)$ *and* $\det \mathbf{J_f}(\mathbf{x}) \neq 0$ *for every* $\mathbf{x} \in \mathbb{R}^n$. *If* $\mathbf{f}^{-1}(K)$ *is compact whenever* K *is compact, prove that* $\mathbf{f}(\mathbb{R}^m) = \mathbb{R}^m$.

Proof. Since $\mathbf{f}$ is continuous on $\mathbb{R}^m$ and $\mathbb{R}^m$ is connected, $\mathbf{f}(\mathbb{R}^m)$ is also connected. We claim that $\mathbf{f}(\mathbb{R}^m)$ is both open and closed in $\mathbb{R}^m$. To this end, let $\mathbf{x} \in \mathbb{R}^m$ and $\mathbf{y} = \mathbf{f}(\mathbf{x}) \in \mathbf{f}(\mathbb{R}^m)$. Since $\det \mathbf{J_f}(\mathbf{x}) \neq 0$, the Inverse Function Theorem implies that there exist open sets $V_{\mathbf{x}}$ and $V_{\mathbf{y}}$ containing $\mathbf{x}$ and $\mathbf{y}$ respectively such that $\mathbf{f} : V_{\mathbf{x}} \to V_{\mathbf{y}}$ is a bijection which gives

$$\mathbf{y} \in V_{\mathbf{y}} = \mathbf{f}(V_{\mathbf{x}}) \subseteq \mathbf{f}(\mathbb{R}^m).$$

In other words, $\mathbf{f}(\mathbb{R}^m)$ is open in $\mathbb{R}^m$.

Next, we prove that $\mathbf{f}(\mathbb{R}^m)$ is closed in $\mathbb{R}^m$. Let $\{\mathbf{y}_n\}$ be a sequence of the set $\mathbf{f}(\mathbb{R}^m)$ and $\mathbf{y}_n \to \mathbf{y} \in \mathbb{R}^m$. Then there is a corresponding sequence $\{\mathbf{x}_n\}$ of $\mathbb{R}^m$ such that

$$\mathbf{f}(\mathbf{x}_n) = \mathbf{y}_n \tag{15.30}$$

for all $n \in \mathbb{N}$. Now the set $K = \{\mathbf{y}_n\} \cup \{\mathbf{y}\}$ is compact, so the hypothesis guarantees that $\mathbf{f}^{-1}(K)$ is also compact and then it is bounded by the Heine-Borel Theorem. By the definition, we have $\{\mathbf{x}_n\} \subseteq \mathbf{f}^{-1}(K)$, so $\{\mathbf{x}_n\}$ is a bounded sequence and the Bolzano-Weierstrass Theorem [25, Problem 5.25] ensures that $\{\mathbf{x}_n\}$ contains a convergent subsequence, namely $\{\mathbf{x}_{n_k}\}$ and $\mathbf{x}_{n_k} \to \mathbf{x} \in \mathbb{R}^m$. Recall that $\mathbf{f}$ is continuous on $\mathbb{R}^m$, so we use the relations (15.30) to conclude that

$$\mathbf{f}(\mathbf{x}) = \lim_{k \to \infty} \mathbf{f}(\mathbf{x}_{n_k}) = \lim_{k \to \infty} \mathbf{y}_{n_k} = \mathbf{y}.$$

Therefore, $\mathbf{f}(\mathbb{R}^m)$ is also closed in $\mathbb{R}^m$ and we have the claim. Since $\det \mathbf{J_f}(\mathbf{x}) \neq 0$, we have $\mathbf{f}(\mathbb{R}^m) \neq \varnothing$ and thus $\mathbf{f}(\mathbb{R}^m) = \mathbb{R}^m$. We complete the proof of the problem. ∎

Problem 15.25

$(\star)$ *Suppose that* $f, g : \mathbb{R} \to \mathbb{R}$ *are continuously differentiable with* $f(0) = 0$ *and* $f'(0) \neq 0$. *Consider the equation*

$$f(x) = tg(x), \tag{15.31}$$

where $t \in \mathbb{R}$. *Prove that there exists a* $\delta > 0$ *such that in* $(-\delta, \delta)$, *there is a unique continuous function* $x(t)$ *satisfying the equation (15.31) and* $x(0) = 0$.

Proof. We define $F : \mathbb{R}^2 \to \mathbb{R}$ by

$$F(x, t) = f(x) - tg(x).$$

Since $f, g, t \in \mathscr{C}'(\mathbb{R})$, we have $F \in \mathscr{C}'(\mathbb{R}^2)$. In addition, we have $F(0, 0) = 0$ and

$$D_1 F(0, 0) = f'(0) - 0 \times g'(0) = f'(0) \neq 0.$$

Hence we follow from the Implicit Function Theorem that one can find a $\delta > 0$ and a unique continuously function $x : (-\delta, \delta) \to \mathbb{R}$ such that

$$x(0) = 0 \quad \text{and} \quad F(x(t), t) = 0$$

which mean that $x(t)$ is a solution of the equation (15.31). Thus we complete the proof of the problem. ∎

Problem 15.26

$(\star)$ Consider the system of equations

$$x^2 + 3y^2 + z^2 - w = 9,$$
$$x^3 + 4y^2 + z + w^2 = 22.$$

Prove that z and w can be written as differentiable functions of x and y around $(1, 1, 1, -4)$.

Proof. We write

$$f_1(x, y, z, w) = x^2 + 3y^2 + z^2 - w - 9 \quad \text{and} \quad f_2(x, y, z, w) = x^3 + 4y^2 + z + w^2 - 22.$$

Next, we define $F : \mathbb{R}^4 \to \mathbb{R}^2$ by

$$\begin{aligned} F(x, y, z, w) &= (f_1(x, y, z, w), f_2(x, y, z, w)) \\ &= (x^2 + 3y^2 + z^2 - w - 9, x^3 + 4y^2 + z + w^2 - 22). \end{aligned}$$

By direct computation, we have $F(1, 1, 1, -4) = (0, 0)$ and

$$\begin{aligned} D_1 f_1 &= 2x, \quad D_2 f_1 = 6y, \quad D_3 f_1 = 2z, \quad D_4 f_1 = -1, \\ D_1 f_2 &= 3x^2, \quad D_2 f_2 = 8y, \quad D_3 f_2 = 1, \quad D_4 f_2 = 2w. \end{aligned}$$

Since all the partial derivatives exist and continuous on $\mathbb{R}^4$, we deduce from Theorem 15.11 that $F \in \mathscr{C}'(\mathbb{R}^4)$. Besides, we know that

$$\det \begin{pmatrix} D_3 f_1(1, 1, 1, -4) & D_4 f_1(1, 1, 1, -4) \\ D_3 f_2(1, 1, 1, -4) & D_4 f_2(1, 1, 1, -4) \end{pmatrix} = \det \begin{pmatrix} 2 & -1 \\ 1 & -8 \end{pmatrix} = -15 \neq 0.$$

Hence it follows from the Implicit Function Theorem that z and w can be written as differentiable functions of x and y around $(1, 1, 1, -4)$. This completes the proof of the problem. ∎

Problem 15.27

$(\star)$ Show that there exist functions $f, g : \mathbb{R}^4 \to \mathbb{R}$ which are functions of x, y, z, w, continuously differentiable in $B_\delta(2, 1, 1, -2)$ for some $\delta > 0$ such that $f(2, 1, 1, -2) = 4$, $g(2, 1, 1, -2) = 3$ and the equations

$$f^2 + g^2 + w^2 = 29 \quad \text{and} \quad \frac{f^2}{x^2} + \frac{g^2}{y^2} + \frac{w^2}{z^2} = 17$$

hold on $B_\delta(2, 1, 1, -2)$.

Proof. Let $F_1(f, g, x, y, z, w) = f^2 + g^2 + w^2 - 29$ and $F_2(f, g, x, y, z, w) = \frac{f^2}{x^2} + \frac{g^2}{y^2} + \frac{w^2}{z^2} - 17$.
Define $\mathbf{F} : \mathbb{R}^6 \to \mathbb{R}^2$ by

$$\mathbf{F}(f, g, x, y, z, w) = (F_1(f, g, x, y, z, w), F_2(f, g, x, y, z, w))$$

$$= \left(f^2 + g^2 + w^2 - 29, \frac{f^2}{x^2} + \frac{g^2}{y^2} + \frac{w^2}{z^2} - 17 \right).$$

It is clear that $\mathbf{F}(4, 3, 2, 1, 1, -2) = (0, 0)$. Furthermore, we have

$$D_1 F_1 = 2f, \quad D_2 F_1 = 2g, \quad D_3 F_1 = 0, \quad D_4 F_1 = 0, \quad D_5 F_1 = 0, \quad D_6 F_1 = 2w,$$

$$D_1 F_2 = \frac{2f}{x^2}, \quad D_2 F_2 = \frac{2g}{y^2}, \quad D_3 F_2 = -\frac{2f^2}{x^3}, \quad D_4 F_2 = -\frac{2g^2}{y^3},$$

$$D_5 F_2 = -\frac{2w^2}{z^3}, \quad D_6 F_2 = \frac{2w}{z^2}.$$

If we take $\delta = \frac{1}{2}$, then any point (x, y, z, w) in $B_{\frac{1}{2}}(2, 1, 1, -2)$ satisfy $xyzw \neq 0$. Otherwise,
assume for example that $y = 0$, then we obtain

$$(x - 2)^2 + (0 - 1)^2 + (z - 1)^2 + (w + 2)^2 < \frac{1}{4}$$

which is a contradiction. In other words, all partial derivatives $D_i F_j$ exist and continuous on
$B_{\frac{1}{2}}(2, 1, 1, -2)$ so that Theorem 15.11 implies $\mathbf{F} \in \mathscr{C}'(B_{\frac{1}{2}}(2, 1, 1, -2))$. Finally, since

$$\det \begin{pmatrix} \dfrac{\partial F_1}{\partial f} & \dfrac{\partial F_1}{\partial g} \\[2mm] \dfrac{\partial F_2}{\partial f} & \dfrac{\partial F_2}{\partial g} \end{pmatrix} = \begin{pmatrix} 2f & 2g \\[2mm] \dfrac{2f}{x^2} & \dfrac{2g}{y^2} \end{pmatrix} = 4fg \left(\frac{1}{y^2} - \frac{1}{x^2} \right)$$

which is nonzero when $f = 4$, $g = 3$, $x = 2$ and $y = 1$, the Implicit Function Theorem ensures
the existence of such functions f and g, completing the proof of the problem. $\blacksquare$

15.5 Higher Order Derivatives

Problem 15.28

(⋆) *Define $f : \mathbb{R}^2 \to \mathbb{R}$ by*

$$f(x, y) = \begin{cases} xy \cdot \dfrac{x^2 - y^2}{x^2 + y^2}, & \text{if } (x, y) \neq (0, 0); \\[3mm] 0, & \text{otherwise.} \end{cases}$$

Prove that $(D_{12}f)(0, 0) \neq (D_{21}f)(0, 0)$.

Proof. Now direct computation gives

$$
(D_1 f)(x,y) = \begin{cases} y \cdot \dfrac{x^2 - y^2}{x^2 + y^2} + \dfrac{4x^2 y^3}{(x^2 + y^2)^2}, & \text{if } (x,y) \neq (0,0); \\[4mm] 0, & \text{otherwise} \end{cases}
$$

and

$$
(D_2 f)(x,y) = \begin{cases} x \cdot \dfrac{x^2 - y^2}{x^2 + y^2} - \dfrac{4x^3 y^2}{(x^2 + y^2)^2}, & \text{if } (x,y) \neq (0,0); \\[4mm] 0, & \text{otherwise.} \end{cases}
$$

Finally, we have

$$
(D_{12} f)(0,0) = \lim_{h \to 0} \frac{(D_2 f)(h,0) - (D_2 f)(0,0)}{h} = \lim_{h \to 0} \frac{1}{h} \times \left(h \cdot \frac{h^2 - 0^2}{h^2 + 0^2} \right) = 1
$$

and

$$
(D_{21} f)(0,0) = \lim_{h \to 0} \frac{(D_1 f)(0,h) - (D_1 f)(0,0)}{h} = \lim_{h \to 0} \frac{1}{h} \times \left(h \cdot \frac{0^2 - h^2}{0^2 + h^2} \right) = -1
$$

which imply that

$$
(D_{12} f)(0,0) \neq (D_{21} f)(0,0),
$$

completing the analysis of the problem. ∎

Problem 15.29

($\star$) *Consider the function $f : \mathbb{R}^2 \to \mathbb{R}$ given by*

$$
f(x,y) = \begin{cases} \dfrac{x^2 y^2}{x^2 + y^2}, & \text{if } (x,y) \neq (0,0); \\[4mm] 0, & \text{otherwise.} \end{cases}
$$

Prove that $(D_{12} f)(0,0) = (D_{21} f)(0,0)$ but $D_{21} f$ is discontinuous at $(0,0)$.

Proof. By the definition, we have

$$
(D_1 f)(x,y) = \begin{cases} \dfrac{2x y^4}{(x^2 + y^2)^2}, & \text{if } (x,y) \neq (0,0); \\[4mm] 0, & \text{otherwise} \end{cases} \tag{15.32}
$$

and

$$
(D_2 f)(x,y) = \begin{cases} \dfrac{2x^4 y}{(x^2 + y^2)^2}, & \text{if } (x,y) \neq (0,0); \\[4mm] 0, & \text{otherwise.} \end{cases}
$$

Furthermore, we have

$$(D_{12}f)(0,0) = \lim_{h \to 0} \frac{(D_2f)(h,0) - (D_2f)(0,0)}{h} = 0$$

and

$$(D_{21}f)(0,0) = \lim_{h \to 0} \frac{(D_1f)(0,h) - (D_1f)(0,0)}{h} = 0$$

so that

$$(D_{12}f)(0,0) = (D_{21}f)(0,0).$$

However, we get from the definition (15.32) that

$$(D_{21}f)(x,y) = \begin{cases} \dfrac{8x^3y^3}{(x^2+y^2)^3}, & \text{if } (x,y) \neq (0,0); \\[2mm] 0, & \text{otherwise.} \end{cases}$$

Therefore, when $x = y = h$, we have

$$\lim_{h \to 0} (D_{21}f)(h,h) = \lim_{h \to 0} \frac{8h^6}{(2h^2)^3} = 1 \neq 0 = (D_{21}f)(0,0).$$

Consequently, $(D_{21}f)(x,y)$ is not continuous at $(0,0)$, completing the proof of the problem. ∎

Remark 15.7

(a) Problem 15.28 tells us that it is *not* always true that $(D_{12}f)(\mathbf{p}) = (D_{21}f)(\mathbf{p})$.

(b) Problem 15.29 is a counterexample that the converse of Theorem 15.12 (Clairaut's Theorem) is false.

Problem 15.30

(★) *Suppose that the third order partial derivatives of the function $f : \mathbb{R}^2 \to \mathbb{R}$ are continuous on $\mathbb{R}^2$. Prove that*

$$D_{122}f = D_{212}f = D_{221}f$$

on $\mathbb{R}^2$.

Proof. If the third order partial derivatives are all continuous on $\mathbb{R}^2$, then so are the second order ones. Hence, by repeated applications of Theorem 15.12 (Clairaut's Theorem), we conclude that

$$D_{221}f = D_2(D_{21}f) = D_2(D_{12}f) = D_{212}f = D_{21}(D_2f) = D_{12}(D_2f) = D_{122}f.$$

This completes the proof of the problem. ∎

Problem 15.31

(★) (★) *Prove Theorem 15.12 (Clairaut's Theorem).*

Proof. If the result holds for each component of $\mathbf{f}$, then it is true for the function $\mathbf{f}$. Therefore, we may assume that we are working with $f : \mathbb{R}^n \to \mathbb{R}$. Furthermore, we prove the theorem for $\mathbf{p} = \mathbf{0}$. Otherwise, we can replace the function $f(\mathbf{x})$ by $f(\mathbf{x} + \mathbf{p})$. The case is trivial if $i = j$, so we assume that $i \neq j$. Suppose that

$$A = (D_{ij}f)(\mathbf{0}) \quad \text{and} \quad B = (D_{ji}f)(\mathbf{0}).$$

Given $\epsilon > 0$. Since $D_{ij}f$ and $D_{ji}f$ are continuous at $\mathbf{0}$, there exists a $\delta > 0$ such that

$$|(D_{ij}f)(\mathbf{x}) - A| < \frac{\epsilon}{2} \quad \text{and} \quad |(D_{ji}f)(\mathbf{x}) - B| < \frac{\epsilon}{2} \tag{15.33}$$

whenever $|\mathbf{x}| < 2\delta$.

Next, we consider

$$J = f(\delta\mathbf{e}_i + \delta\mathbf{e}_j) - f(\delta\mathbf{e}_i) - [f(\delta\mathbf{e}_j) - f(\mathbf{0})]. \tag{15.34}$$

Applying the Second Fundamental Theorem of Calculus [25, p. 161] to the variable in $\mathbf{e}_i$, we see that

$$f(\delta\mathbf{e}_i + \delta\mathbf{e}_j) - f(\delta\mathbf{e}_j) = \int_0^\delta (D_i f)(x_i\mathbf{e}_i + \delta\mathbf{e}_j)\, \mathrm{d}x_i \tag{15.35}$$

and

$$f(\delta\mathbf{e}_i) - f(\mathbf{0}) = \int_0^\delta (D_i f)(x_i\mathbf{e}_i)\, \mathrm{d}x_i. \tag{15.36}$$

Now we substitute the expressions (15.35) and (15.36) into the definition (15.34) to get

$$J = \int_0^\delta \left[(D_i f)(x_i\mathbf{e}_i + \delta\mathbf{e}_j) - (D_i f)(x_i\mathbf{e}_i) \right] \mathrm{d}x_i. \tag{15.37}$$

By applying the Mean Value Theorem for Derivatives [25, p. 129] to the differentiable function $D_i f$ with respect to the variable in $\mathbf{e}_j$, we obtain

$$(D_i f)(x_i\mathbf{e}_i + \delta\mathbf{e}_j) - (D_i f)(x_i\mathbf{e}_i) = (\delta - 0)[D_j(D_i f)](x_i\mathbf{e}_i + x_j\mathbf{e}_j)$$
$$= \delta(D_{ji}f)(x_i\mathbf{e}_i + x_j\mathbf{e}_j)$$

for some $x_j \in (0, \delta)$. Since $x_i, x_j \in (0, \delta)$, we know that $|x_i\mathbf{e}_i + x_j\mathbf{e}_j| < 2\delta$ so that the first inequality (15.33) gives

$$|(D_i f)(x_i\mathbf{e}_i + \delta\mathbf{e}_j) - (D_i f)(x_i\mathbf{e}_i) - \delta A| = |\delta(D_{ji}f)(x_i\mathbf{e}_i + x_j\mathbf{e}_j) - \delta A|$$
$$= \delta|(D_{ji}f)(x_i\mathbf{e}_i + x_j\mathbf{e}_j) - A|$$
$$< \frac{\delta\epsilon}{2}$$

or equivalently

$$-\frac{\delta\epsilon}{2} < (D_i f)(x_i\mathbf{e}_i + \delta\mathbf{e}_j) - (D_i f)(x_i\mathbf{e}_i) - \delta A < \frac{\delta\epsilon}{2}. \tag{15.38}$$

Here we integrable each part of the inequalities (15.38) from 0 to δ and then use the representation (15.37) to get

$$|J - \delta^2 A| < \frac{\delta^2\epsilon}{2}. \tag{15.39}$$

Similarly, we can verify that

$$|J - \delta^2 B| < \frac{\delta^2\epsilon}{2}. \tag{15.40}$$

Finally, we deduce immediately from the inequalities (15.39) and (15.40) that

$$|\delta^2 A - \delta^2 B| \leq |\delta^2 A - J| + |J - \delta^2 B| < \frac{\delta^2 \epsilon}{2} + \frac{\delta^2 \epsilon}{2} = \delta^2 \epsilon$$

which implies that

$$|A - B| < \epsilon.$$

Since ϵ is arbitrary, we conclude that

$$A = B$$

which completes the proof of the problem.   ∎

CHAPTER $\boldsymbol{16}$

Integral Calculus of Functions of Several Variables

16.1 Fundamental Concepts

In this chapter, we study the Riemann integral of real-valued functions of several real variables. This is a direct generalization of the Riemann integral of functions of one variable which we have investigated in [25, Chap. 9]. The main references of this chapter are [2, Chap. 14], [21, Chap. 1], [23, §1.1], [24, Chap. 12] and [28, Chap. 11].

16.1.1 The Measure of Intervals in $\mathbb{R}^n$

Definition 16.1 (Intervals in $\mathbb{R}^n$). *For $1 \leq k \leq n$, consider $-\infty < a_k \leq b_k < \infty$. Then each $I_k = [a_k, b_k]$ is an (closed) interval in $\mathbb{R}$. We call I an (closed) n-**dimensional interval in $\mathbb{R}^n$** if I has the form*

$$I = I_1 \times I_2 \times \cdots \times I_n = \{(x_1, \ldots, x_n) \,|\, x_k \in I_k \text{ for } 1 \leq k \leq n\}. \tag{16.1}$$

Definition 16.2 (The Measure of Intervals in $\mathbb{R}^n$). *The **measure** or **volume** of the interval I in the form (16.1), denoted by $\mu(I)$, is defined to be*

$$\mu(I) = \mu(I_1) \times \mu(I_2) \times \cdots \times \mu(I_n) = \prod_{k=1}^{n}(b_k - a_k). \tag{16.2}$$

When $n = 1, 2$ and 3, the value (16.2) corresponds to the **length**, the **area** and the **volume of** I.[a]

Theorem 16.3. *Suppose that $I, \mathcal{I}_1, \mathcal{I}_2, \ldots, \mathcal{I}_m$ are intervals in $\mathbb{R}^n$ having the form (16.1). Then the measure of the interval I in $\mathbb{R}^n$ has the following properties:*

---

[a]It is easily seen that $\mu(I) = 0$ if $\mu(I_k) = 0$ for some k. Furthermore, if each I_k is an open interval in $\mathbb{R}$, then the form (16.1) gives us an open interval I in $\mathbb{R}^n$ and it has the *same* measure as its corresponding closed interval.

(a) If $I \subseteq \mathcal{I}_1 \cup \mathcal{I}_2 \cup \cdots \cup \mathcal{I}_m$, then we have

$$\mu(I) \leq \sum_{k=1}^{m} \mu(\mathcal{I}_k).$$

(b) If $I = \bigcup_{k=1}^{m} \mathcal{I}_k$ and no two of $\mathcal{I}_1, \mathcal{I}_2, \ldots, \mathcal{I}_m$ have a common interior point (i.e., $\mathcal{I}_s^{\circ} \cap \mathcal{I}_t^{\circ} = \varnothing$ if $s \neq t$), then we have

$$\mu(I) = \sum_{k=1}^{m} \mu(\mathcal{I}_k).$$

Readers should discover the similarity between Theorem 16.3 and the inequality (12.1) and Theorem 12.11 (Countably Additivity).

16.1.2 The Riemann Integral in $\mathbb{R}^n$

Definition 16.4 (Partitions of I). *Suppose that $I = I_1 \times \cdots \times I_n$ is an interval in $\mathbb{R}^n$. Let P_k be a partition of I_k.[b] Then the product*

$$P = P_1 \times P_2 \times \cdots \times P_n \tag{16.3}$$

is called a **partition of** I*.[c] A partition P' of I is said to be* **finer than** *the partition P of I if $P \subseteq P'$.*

It is clear that if P_k makes I_k into m_k subintervals, then the partition (16.3) decomposes I into a union of $(m_1 \times m_2 \times \cdots \times m_n)$ **subintervals in** $\mathbb{R}^n$. See Figure 16.1 for an example of a collection of subintervals in $\mathbb{R}^2$, where each small rectangle is a subinterval of the large rectangle.

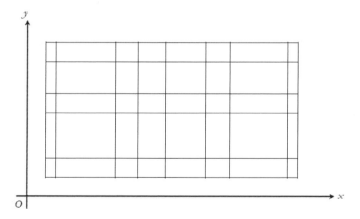

Figure 16.1: The subinterval in $\mathbb{R}^2$.

---

[b]For the definition of a partition of $[a, b]$, see [25, p. 157].
[c]Some books call it a **grid of** I.

Definition 16.5 (Riemann Sum). *Suppose that I is an interval in $\mathbb{R}^n$ and $f : I \to \mathbb{R}$ is a bounded function. Let P be a partition of I into m subintervals $I_1, I_2, \ldots, I_m$ and $\mathbf{t}_k \in I_k$. Then the sum*

$$S(P, f) = \sum_{k=1}^{m} f(\mathbf{t}_k)\mu(I_k)$$

*is called a **Riemann sum**.*

Definition 16.6 (Riemann Integrable on I). *The function $f : I \subseteq \mathbb{R}^n \to \mathbb{R}$ is called **Riemann integrable on** I, denoted by $f \in \mathscr{R}$ on I, if there corresponds a **unique** real number A whenever for every $\epsilon > 0$, there exists a partition P_ϵ of I such that*

$$|S(P, f) - A| < \epsilon$$

for every partition P finer than P_ϵ. When this number exists, we write

$$A = \int_I f \, \mathrm{d}\mathbf{x} = \int_I f(x_1, x_2, \ldots, x_n) \, \mathrm{d}x_1 \, \mathrm{d}x_2 \cdots \mathrm{d}x_n. \tag{16.4}$$

Remark 16.1

When $n = 2$ and $n = 3$ in the integral (16.4), it is our usual **double** and **triple integral** and they are denoted by

$$\iint_I f(x, y) \, \mathrm{d}x \, \mathrm{d}y \quad \text{and} \quad \iiint_I f(x, y, z) \, \mathrm{d}x \, \mathrm{d}y \, \mathrm{d}z.$$

respectively.

Definition 16.7 (Lower and Upper Riemann Integrals). *Suppose that I is an interval in $\mathbb{R}^n$ and $f : I \to \mathbb{R}$ is a bounded function. Let P be a partition of I into p subintervals $I_1, I_2, \ldots, I_p$ and let*

$$m_k(f) = \inf\{f(\mathbf{x}) \mid \mathbf{x} \in I_k\} \quad \text{and} \quad M_k(f) = \sup\{f(\mathbf{x}) \mid \mathbf{x} \in I_k\},$$

where $k = 1, 2, \ldots, p$. The two sums

$$L(P, f) = \sum_{k=1}^{p} m_k(f)\mu(I_k) \quad \text{and} \quad U(P, f) = \sum_{k=1}^{p} M_k(f)\mu(I_k)$$

*are called the **lower** and **upper Riemann sums** respectively. Then the **lower** and **upper Riemann integrals** of f on I are defined by*

$$\underline{\int}_I f \, \mathrm{d}\mathbf{x} = \sup\{L(P, f) \mid P \text{ is a partition of } I\}$$

and

$$\overline{\int}_I f \, \mathrm{d}\mathbf{x} = \inf\{U(P, f) \mid P \text{ is a partition of } I\}$$

respectively.

16.1.3 Criteria for Integrability on Intervals

Theorem 16.8 (The Riemann Integrability Condition). *We have $f \in \mathscr{R}$ on the interval I in $\mathbb{R}^n$ if and only if for every $\epsilon > 0$, there exists a partition P_ϵ such that*

$$U(P_\epsilon, f) - L(P_\epsilon, f) < \epsilon. \tag{16.5}$$

In addition, the inequality (16.5) holds for every refinement of P_ϵ and the inequality (16.5) is also equivalent to the condition that

$$\underline{\int_I} f \, d\mathbf{x} = \overline{\int_I} f \, d\mathbf{x} = \int_I f \, d\mathbf{x}.$$

Definition 16.9 (n-measure zero). *A subset $S \subseteq \mathbb{R}^n$ is of n-**measure zero**, i.e., $\mu(S) = 0$, provided for every $\epsilon > 0$, there exists a countable collection of n-dimensional intervals $I_1, I_2, \ldots$ such that*

$$S \subseteq \bigcup_{k=1}^\infty I_k \quad \text{and} \quad \sum_{k=1}^\infty \mu(I_k) < \epsilon.$$

Theorem 16.10 (The Lebesgue's Integrability Condition). *Suppose that I is an interval in $\mathbb{R}^n$ and $f : I \to \mathbb{R}$ is a bounded function. Then we have $f \in \mathscr{R}$ on I if and only if the set of discontinuities of f in I has n-measure zero. Particularly, we have $f \in \mathscr{R}$ on I for every continuous function $f : I \to \mathbb{R}$.*

Remark 16.2

(a) In particular, if $\mu(S_k) = 0$ for $k = 1, 2, \ldots$ and $S \subseteq S_1 \cup S_2 \cup \cdots$, then we have $\mu(S) = 0$.

(b) It is easy to see from Definition 16.7 (Lower and Upper Riemann Integrals) that if $f(\mathbf{x}) = 0$ on I, then $L(P, f) = U(P, f) = 0$ for every partition P of I. Hence Theorem 16.8 (The Riemann Integrability Condition) implies that $f \in \mathscr{R}$ on I and

$$\int_I f \, d\mathbf{x} = \int_I 0 \, d\mathbf{x} = 0. \tag{16.6}$$

(c) We suggest the reader to compare our Theorems 16.8 (The Riemann Integrability Condition) and 16.10 (The Lebesgue's Integrability Condition) with [25, Theorems 9.2 and 9.4, p. 159] respectively.

16.1.4 Jordan Measurable Sets in $\mathbb{R}^n$

Roughly speaking, the **Jordan measure** (or **Jordan content**) in $\mathbb{R}^n$ is an extension of the notion of size (length, area and volume) to more complicated shapes other than triangles or disks.

Definition 16.11 (Jordan Measure). *Suppose that $E \subseteq I$, where I is an interval in the form (16.1). We define the **outer Jordan measure** of E to be*

$$J^*(E) = \inf \sum_{k=1}^N \mu(I_k),$$

where the infimum runs through over all finite coverings $E \subseteq \bigcup_{k=1}^{N} I_k$ by intervals in $\mathbb{R}^n$. Similarly, the **inner Jordan measure** of E is defined to be

$$J_*(E) = \sup \sum_{k=1}^{N} \mu(I_k),$$

where the supremum runs through over all unions of finite intervals in $\mathbb{R}^n$ which are contained in E. In the case that

$$J^*(E) = J_*(E),$$

the set E is said to be **Jordan measurable** and we denote this common value by $J(E)$ which is called the **Jordan measure** or **Jordan content** of E.

For examples, every finite subset of $\mathbb{R}^n$ is Jordan measurable and all open and closed balls in $\mathbb{R}^n$ are Jordan measurable too. In Figure 16.2, the sum of the areas of the orange rectangles is an *inside* approximation of the area of the Jordan measurable set in $\mathbb{R}^2$. Similarly, the sum of the areas of the green rectangles is an *outside* approximation of the area of the Jordan measurable set in $\mathbb{R}^2$.

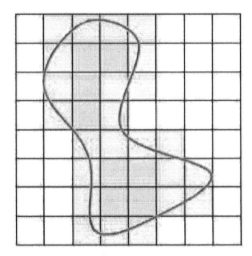

Figure 16.2: The outer and the inner Jordan measures.

Remark 16.3

Readers should compare the similarities between Definition 16.11 (Jordan Measure) and Definition 12.2 (Lebesgue Outer Measure).

Definition 16.12 (Boundary Points). *We call that a point* $\mathbf{x} \in E \subseteq \mathbb{R}^n$ *is a* **boundary point** *of* E *if*

$$N_r(\mathbf{x}) \cap E \neq \varnothing \quad and \quad N_r(\mathbf{x}) \cap (\mathbb{R}^n \setminus E) \neq \varnothing$$

for every $r > 0$. *The set of all boundary points of E is called the* **boundary** *of E and it is denoted by ∂E.*

Theorem 16.13. *Suppose that E is a bounded set of $\mathbb{R}^n$. Then we have*

$$J^*(\partial E) = J^*(E) - J_*(E).$$

In addition, E is Jordan measurable if and only if $J(\partial E) = 0$.

It is easy to see that if $J(E) = 0$, then $\mu(E) = 0$. However, the converse is not true. See Problem 16.2 for proofs of these. The following theorem tells us some basic properties of Jordan measurable sets.

Theorem 16.14 (Properties of Jordan Measurable Sets). *Suppose that E and F are Jordan measurable. Then $E \cup F$, $E \cap F$ and $E \setminus F$ are also Jordan measurable. Furthermore, we have*

(a) $J(E \cup F) = J(E) + J(F) - J(E \cap F)$.

(b) *if $E \cap F = \varnothing$, then $J(E \cup F) = J(E) + J(F)$.*

(c) *if $F \subseteq E$, then $J(E \setminus F) = J(E) - J(F)$.*

▌ 16.1.5   Integration on Jordan Measurable Sets

Definition 16.15 (Riemann Integrable on E). *Suppose that E is a Jordan measurable set of $\mathbb{R}^n$, f is bounded on E and I is an interval in the form (16.1) containing E. Define $g : I \to \mathbb{R}$ by*

$$g(\mathbf{x}) = \begin{cases} f(\mathbf{x}), & \text{if } \mathbf{x} \in E; \\ 0, & \text{if } \mathbf{x} \in I \setminus E. \end{cases}$$

We say that $f \in \mathscr{R}$ on E if and only if $g \in \mathscr{R}$ on I and in this case, we have

$$\int_E f \, d\mathbf{x} = \int_I g \, d\mathbf{x}.$$

Theorem 16.16. *Suppose that E is a Jordan measurable set of $\mathbb{R}^n$. Then we have $f \in \mathscr{R}$ on E if and only if the set of discontinuous points of f in E has n-**measure** 0. In particular, $f \in \mathscr{R}$ on E for every continuous function $f : E \to \mathbb{R}$.*

Theorem 16.17 (Properties of Integration on E). *Suppose that E is a Jordan measurable set in $\mathbb{R}^n$, $f, g : E \to \mathbb{R}$ and $\alpha \in \mathbb{R}$.*

(a) *If $f, g \in \mathscr{R}$ on E, then so are $f + g$ and αf. In fact, we have*

$$\int_E (f + g) \, d\mathbf{x} = \int_E f \, d\mathbf{x} + \int_E g \, d\mathbf{x}$$

and

$$\int_E \alpha f \, d\mathbf{x} = \alpha \int_E f \, d\mathbf{x}.$$

(b) If $f, g \in \mathscr{R}$ on E and $f(\mathbf{x}) \le g(\mathbf{x})$ for all $\mathbf{x} \in E$, then we have

$$\int_E f \, \mathrm{d}\mathbf{x} \le \int_E g \, \mathrm{d}\mathbf{x}.$$

(c) If A and B are Jordan measurable sets satisfying $A \cap B = \varnothing$ and $E = A \cup B$ and $f \in \mathscr{R}$ on E, then we have $f \in \mathscr{R}$ on A and on B and furthermore,

$$\int_E f \, \mathrm{d}\mathbf{x} = \int_A f \, \mathrm{d}\mathbf{x} + \int_B f \, \mathrm{d}\mathbf{x}.$$

The following famous theorem suggests a practical way to evaluate double integrals on a rectangle in terms of singles integrals. For multiple integrals on an interval in $\mathbb{R}^n$, please read Problem 16.11.

Fubini's Theorem. *Suppose that $R = [a,b] \times [c,d]$ and $f : R \to \mathbb{R}$ is well-defined. Furthermore, suppose that for each $x \in [a,b]$, $g(y) = f(x, \cdot) \in \mathscr{R}$ on $[c,d]$ and for each $y \in [c,d]$, $h(x) = f(\cdot, y) \in \mathscr{R}$ on $[a,b]$. If $f \in \mathscr{R}$ on R, then we have*

$$\int_R f(x,y) \, \mathrm{d}(x,y) = \iint_R f(x,y) \, \mathrm{d}x \, \mathrm{d}y = \int_a^b \Big(\int_c^d f(x,y) \, \mathrm{d}y \Big) \, \mathrm{d}x = \int_c^d \Big(\int_a^b f(x,y) \, \mathrm{d}x \Big) \, \mathrm{d}y.$$

For evaluation of integrals on a general Jordan measurable set in terms of iterated integrals, we consider the concept of a projectable region first. We call a set $E \subseteq \mathbb{R}^n$ a **projectable region** if there exists a closed Jordan measurable set $F \subseteq \mathbb{R}^{n-1}$, $k \in \{1, 2, \ldots, n\}$ and continuous functions $\phi, \varphi : F \subseteq \mathbb{R}^{n-1} \to \mathbb{R}$ such that

$$E = \{ \mathbf{x} \in \mathbb{R}^n \, | \, \widehat{\mathbf{x}}_k \in F \text{ and } \phi(\widehat{\mathbf{x}}_k) \le x_k \le \varphi(\widehat{\mathbf{x}}_k) \}, \tag{16.7}$$

where $\widehat{\mathbf{x}}_k = (x_1, \ldots, x_{k-1}, x_{k+1}, \ldots, x_n)$. Then we have

Theorem 16.18. *Suppose that E is a projectable region in $\mathbb{R}^n$ with k, F, ϕ and φ as given in the definition (16.7). Then E is a Jordan measurable set and if $f \in \mathscr{R}$ on E, then*

$$\int_E f \, \mathrm{d}\mathbf{x} = \int_F \Big(\int_{\phi(\widehat{\mathbf{x}}_k)}^{\varphi(\widehat{\mathbf{x}}_k)} f(\widehat{\mathbf{x}}_k) \, \mathrm{d}x_k \Big) \, \mathrm{d}\widehat{\mathbf{x}}_k.$$

▌16.1.6 Two Important Theorems

In [25, Chap. 9], we study the Mean Value Theorems for Integrals and the Change of Variables Theorem in the case of a single variable. Here we present their multi-dimensional versions as follows:

The Mean Value Theorem for Multiple Integrals. *Suppose that E is a Jordan measurable set of $\mathbb{R}^n$. Let $f, g \in \mathscr{R}$ on E and $g(\mathbf{x}) \ge 0$ on E. Denote*

$$M = \sup\{f(\mathbf{x}) \, | \, \mathbf{x} \in E\} \quad \text{and} \quad m = \inf\{f(\mathbf{x}) \, | \, \mathbf{x} \in E\}.$$

Then there is a number $\lambda \in [m, M]$ such that

$$\int_E fg \, \mathrm{d}\mathbf{x} = \lambda \int_E g \, \mathrm{d}\mathbf{x}.$$

Particularly, we have

$$mJ(E) \le \int_E f \, \mathrm{d}\mathbf{x} \le MJ(E).$$

The Change of Variables Theorem. *Suppose that V is open in $\mathbb{R}^n$, $\phi : V \to \mathbb{R}^n$ is continuously differentiable and injective on V. If $\det \mathbf{J}_\phi(\mathbf{x}) \neq 0$ for all $\mathbf{x} \in V$, E is a Jordan measurable set with $\overline{E} \subseteq V$, $f \circ \phi \in \mathscr{R}$ on E and $f \in \mathscr{R}$ on $\phi(E)$, then we have*

$$\int_{\phi(E)} f(\mathbf{y})\,\mathrm{d}\mathbf{y} = \int_E f(\phi(\mathbf{x})) \cdot |\det \mathbf{J}_\phi(\mathbf{x})|\,\mathrm{d}\mathbf{x}. \tag{16.8}$$

Recall that **polar coordinates** in $\mathbb{R}^2$ have the form

$$x = r\cos\theta \quad \text{and} \quad y = r\sin\theta,$$

where r is the distance between (x,y) and the origin and θ is the angle between the positive x-axis and the line connecting (x,y) and $(0,0)$. Set $\phi(r,\theta) = (r\cos\theta, r\sin\theta)$ so that

$$\det \mathbf{J}_\phi((r,\theta)) = \det \begin{pmatrix} \cos\theta & -r\sin\theta \\ \sin\theta & r\cos\theta \end{pmatrix} = r.$$

Obviously, ϕ is continuously differentiable on any open set V in $\mathbb{R}^2$ by Theorem 15.11. Furthermore, ϕ is injective and $\det \mathbf{J}_\phi(r,\theta) \neq 0$ on V if V does not intersect the set $\{(0,\theta)\,|\,\theta \in \mathbb{R}\}$. Hence the formula (16.8) becomes

$$\int_{\phi(E)} f(x,y)\,\mathrm{d}x\,\mathrm{d}y = \int_E f(r\cos\theta, r\sin\theta)r\,\mathrm{d}r\,\mathrm{d}\theta. \tag{16.9}$$

Remark 16.4

(a) There are two common, helpful and practical changes of variables in $\mathbb{R}^3$. They are the **cylindrical coordinates** and the **spherical coordinates** which are given by

$$x = r\cos\theta, \quad y = r\sin\theta, \quad z = z$$

and

$$x = r\sin\varphi\cos\theta, \quad y = r\sin\varphi\sin\theta, \quad z = r\cos\theta$$

respectively.

(b) In [24, p. 426], Wade points out that the formula (16.9) holds even though ϕ is not injective or $\det \mathbf{J}_\phi((r,\theta)) = 0$ in the whole $r\theta$-plane. Similar situations happen for the cylindrical coordinates and the spherical coordinates.

(c) Wade calls this Change of Variables Theorem a global version which can be shown by a local version of the theorem, see [24, Lemma 12.45, p. 424].

16.2 Jordan Measurable Sets

Problem 16.1

⊛ Prove that $J^*(E) = J^*(\overline{E})$ for every $E \subseteq \mathbb{R}$.

Proof. Since $E \subseteq \overline{E}$, we always have

$$E \subseteq \overline{E} \subseteq \bigcup_{k=1}^{N} I_k$$

so that $J^*(E) \le J^*(\overline{E})$. To prove the reverse direction, it suffices to show that if $E \subseteq \bigcup_{k=1}^{N} I_k$,

then $\overline{E} \subseteq \bigcup_{k=1}^{N} I_k$. To this end, let $I = \bigcup_{k=1}^{N} I_k$. Assume that $p \in \overline{E}$ but $p \notin I$. Particularly, $p \notin E$.
Since $\overline{E} = E \cup E'$, we have $p \in E'$ which implies that $(p - \epsilon, p + \epsilon) \cap E \ne \varnothing$ for every $\epsilon > 0$ and
then

$$(p - \epsilon, p + \epsilon) \cap I \ne \varnothing$$

for every $\epsilon > 0$. Thus p is a limit point of I. By Definition 16.1 (Intervals in $\mathbb{R}^n$), I is closed in
$\mathbb{R}^n$, so $p \in I$ which is a contradiction. Therefore, it is true that $\overline{E} \subseteq I$ and then $J^*(\overline{E}) \le J^*(E)$.
Consequently, we get the desired result that

$$J^*(E) = J^*(\overline{E}).$$

This completes the proof of the problem. ∎

Problem 16.2

(⋆) Prove that if $J(E) = 0$, then $\mu(E) = 0$. Show also that the converse is not true.

Proof. Suppose that $J(E) = 0$. Given $\epsilon > 0$. By Definition 16.11 (Jordan Measure), there is a
finite collection of intervals $\{I_1, I_2, \ldots, I_N\}$ whose union covers E such that

$$\sum_{k=1}^{N} \mu(I_k) < \epsilon. \tag{16.10}$$

If we consider the countable collection $\{I_1, I_2, \ldots, I_N, I_{N+1}, \ldots\}$, where $I_n = \varnothing$ for all $n \ge N+1$,
then its union also covers E. Now the estimate (16.10) and the fact $\mu(\varnothing) = 0$ certainly give

$$\sum_{k=1}^{\infty} \mu(I_k) < \epsilon.$$

By Theorem 16.3(a), we see that $\mu(E) = 0$.

Let $F = \mathbb{Q} \cap [0, 1]$. We know from Problem 12.3 and Theorem 12.5(a) (Properties of Measurable Sets) that $m(F) = 0$. Since F is dense in $[0, 1]$, we observe from Problem 16.1 that

$$J^*(F) = J^*(\overline{F}) = J^*([0, 1]) = 1.$$

However, the density of irrationals in $[0, 1]$ implies that F has *only* empty interval so that
$J_*(F) = 0$. Since $J_*(F) \ne J^*(F)$, Definition 16.11 (Jordan Measure) shows that F is not Jordan
measurable. This completes the proof of the problem. ∎

Remark 16.5

(a) By Problem 16.2, the same result (16.6) holds for any bounded f if $J(E) = 0$.

(b) In addition, we recall that a rational $q \in [0, 1]$ is Jordan measurable because $\{q\}$ is a finite set, but their union is $F = \mathbb{Q} \cap [0, 1]$ in Problem 16.2 is not Jordan measurable. In other words, this counterexample shows that countable union of Jordan measurable sets is *not* necessarily Jordan measurable.

Problem 16.3

$(\star)$ If $E \subseteq F \subseteq \mathbb{R}^n$, prove that $J^*(E) \leq J^*(F)$.

Proof. Suppose that $\{I_1, I_2, \ldots, I_N\}$ is a finite collection of intervals in $\mathbb{R}^n$ covering F. Then we have

$$E \subseteq F \subseteq \bigcup_{k=1}^{N} I_k. \tag{16.11}$$

By Definition 16.11 (Jordan Measure), we have

$$J^*(E) \leq \sum_{k=1}^{N} \mu(I_k)$$

for every finite collection of intervals satisfying the set relations (16.11). By Definition 16.11 (Jordan Measure) again, we conclude that $J^*(E) \leq J^*(F)$ which completes the proof of the problem. ∎

Problem 16.4

$(\star)$ Suppose that E is a Jordan measurable subset of $\mathbb{R}^n$ and $J(E) = 0$. If $F \subseteq E$, prove that F is also Jordan measurable and $J(F) = 0$.

Proof. Since $J(E) = 0$, we have $J^*(E) = 0$ by Definition 16.11 (Jordan Measure). By Problem 16.3, we know that $J^*(F) = 0$. By the definition again, we always have

$$J_*(F) \leq J^*(F).$$

Thus we get $J_*(F) = J^*(F) = 0$ and the definition again shows that F is Jordan measurable and $J(F) = 0$. We complete the proof of the problem. ∎

Problem 16.5

$(\star)$ Let $E \subseteq \mathbb{R}$, prove that

$$J^*(E) = \inf J^*(V),$$

where the infimum takes over all open sets V containing E.

Proof. Let $S = \{J^*(V) \,|\, V \text{ is an open set containing } E\}$. If $J^*(E) = \infty$, then Problem 16.4 implies that $J^*(V) = \infty$ and our result follows. Therefore, we may assume that $J^*(E) < \infty$. By Problem 16.4 again, $J^*(E)$ is a lower bound of S. Now we want to show that for every $\epsilon > 0$, there is an open sets V containing E such that

$$J^*(V) \le J^*(E) + \epsilon. \tag{16.12}$$

To see this, given $\epsilon > 0$, then the definition of the infimum implies that there exists a finite collection of intervals $\{I_1, I_2, \ldots, I_N\}$ in $\mathbb{R}^n$ covering E such that

$$\sum_{k=1}^{N} \mu(I_k) \le J^*(E) + \frac{\epsilon}{2}. \tag{16.13}$$

Recall from Definition 16.1 (Intervals in $\mathbb{R}^n$) that there are $a_{k1}, b_{k1}, a_{k2}, b_{k2}, \ldots, a_{kn}, b_{kn}$ such that

$$I_k = [a_{k1}, b_{k1}] \times [a_{k2}, b_{k2}] \times \cdots \times [a_{kn}, b_{kn}],$$

where $k = 1, 2, \ldots, N$. Let $\delta > 0$ and we consider the open intervals

$$V_k = (a_{k1} - \delta, b_{k1} + \delta) \times (a_{k2} - \delta, b_{k2} + \delta) \times \cdots \times (a_{kn} - \delta, b_{kn} + \delta).$$

Then we have $I_k \subseteq V_k$ and

$$\mu(V_k) = \prod_{j=1}^{N} (b_{kj} - a_{kj} + 2\delta)$$

by the definition (16.2). Now we can make δ as small as possible so that

$$\mu(V_k) \le \mu(I_k) + \frac{\epsilon}{2^{k+1}}. \tag{16.14}$$

If we define $V = \bigcup_{k=1}^{N} V_k$, then it is open in $\mathbb{R}^n$ and

$$E \subseteq \bigcup_{k=1}^{N} I_k \subseteq V,$$

so we follow from Definition 16.11 (Jordan Measure) and the inequalities (16.14) that

$$J^*(V) \le \sum_{k=1}^{N} \mu(V_k) \le \sum_{k=1}^{N} \left[\mu(I_k) + \frac{\epsilon}{2^{k+1}} \right] = \sum_{k=1}^{N} \mu(I_k) + \frac{\epsilon}{2}. \tag{16.15}$$

Combining the inequalities (16.13) and (16.15), we gain

$$J^*(V) \le J^*(E) + \epsilon$$

which is exactly the required result (16.12). Hence we obtain $J^*(E) = \inf J^*(V)$ and this completes the proof of the problem. ∎

Problem 16.6

(★) *Prove that there exists a non-Jordan measurable subset of* $S = [0,1] \times [0,1]$.

Proof. We consider the subset

$$E = \{(x, y) \,|\, x, y \in \mathbb{Q} \cap [0, 1]\}.$$

We note that an interval in $\mathbb{R}^2$ is actually a rectangle. On the one hand, it is clear that there is *no* rectangle contained in E because a rectangle must contain a point with irrational coordinates. Thus it means that

$$J_*(E) = 0.$$

On the other hand, let R be an interval contained in S. Then the density of $\mathbb{Q}$ shows that

$$R \cap E \neq \varnothing. \qquad (16.16)$$

This observation means that if $E \subseteq \bigcup\limits_{k=1}^{N} I_k$, then we have

$$\bigcup_{k=1}^{N} I_k = S.$$

Otherwise, since all I_k are rectangles, one can find a rectangle R such that

$$R \subseteq S \setminus \bigcup_{k=1}^{N} I_k$$

and then $R \cap E = \varnothing$ which contradicts the observation (16.16). Therefore, we have

$$J^*(E) = \mu(S) = 1.$$

Consequently, $J^*(E) \neq J_*(E)$ and it follows from Definition 16.11 (Jordan Measure) that E is not Jordan measurable set. This ends the analysis of the problem. ∎

16.3 Integration on $\mathbb{R}^n$

Problem 16.7

(⋆) *Suppose that E is a compact Jordan measurable set in $\mathbb{R}^n$. Prove that*

$$J(E) = \int_E d\mathbf{x}.$$

Proof. Let I be an interval containing E in the form (16.1). By the definition, we have

$$\chi_E(\mathbf{x}) = \begin{cases} 1, & \text{if } \mathbf{x} \in E; \\ 0, & \text{if } \mathbf{x} \in I \setminus E. \end{cases}$$

Clearly, the set of discontinues points of χ_E in I are exactly ∂E. Since E is Jordan measurable, it follows from Theorem 16.13 that $J(\partial E) = 0$. By Problem 16.2, we have $\mu(\partial E) = 0$ and then Theorem 16.10 (The Lebegus's Integrability Condition) ensures that $\chi_E \in \mathscr{R}$ on I. By Definition 16.15 (Riemann Integrable on E), we conclude that

$$\int_E d\mathbf{x}$$

exists and then Theorem 16.8 (The Riemann Integrability Condition) guarantees that

$$\overline{\int}_E d\mathbf{x} = \underline{\int}_E d\mathbf{x} = \int_E d\mathbf{x}. \tag{16.17}$$

Next, we let P be a partition of I into subintervals $I_1, I_2, \ldots, I_p$ and

$$S = \{k \in \{1, 2, \ldots, p\} \,|\, I_k \cap E \neq \varnothing\}.$$

Therefore, we obtain

$$E \subseteq \bigcup_{k \in S} I_k$$

and

$$M_k(\chi_E) = \sup\{\chi_E(\mathbf{x}) \,|\, \mathbf{x} \in I_k\} = \begin{cases} 1, & \text{if } k \in S; \\ 0, & \text{if } k \notin S. \end{cases}$$

By Definition 16.7 (Lower and Upper Riemann Integrals) and then Definition 16.11 (Jordan Measure), we see that

$$U(P, \chi_E) = \sum_{k=1}^{p} M_k(\chi_E)\mu(I_k) = \sum_{k \in S} \mu(I_k)$$

which implies

$$\overline{\int}_I \chi_E \, d\mathbf{x} = \inf\{U(P, \chi_E) \,|\, P \text{ is a partition of } I\}$$

$$= \inf\left\{\sum_{k \in S} \mu(I_k) \,\Big|\, E \subseteq \bigcup_{k \in S} I_k\right\}$$

$$= J^*(E). \tag{16.18}$$

Since E is Jordan measurable, the expression (16.18) reduces to

$$\overline{\int}_I \chi_E \, d\mathbf{x} = J(E). \tag{16.19}$$

By combining the expressions (16.17) and (16.19), we establish that

$$J(E) = \overline{\int}_I \chi_E \, d\mathbf{x} = \overline{\int}_E d\mathbf{x} = \int_E d\mathbf{x},$$

completing the proof of the problem. ∎

Remark 16.6

In fact, the condition that E is compact can be dropped in Problem 16.7.

Problem 16.8

(⋆) *Prove Theorem 16.16.*

Proof. Let I be an interval containing E. Define

$$
g(\mathbf{x}) = \begin{cases} f(\mathbf{x}), & \text{if } \mathbf{x} \in E; \\[2mm] 0, & \text{if } \mathbf{x} \in I \setminus E. \end{cases}
$$

Let $D_f(E)$ be the set of all discontinuities of f on E. By Theorem 16.10 (The Lebesgue's Integrability Condition) or Theorem 16.16, $g \in \mathscr{R}$ on I if and only if $\mu(D_g(I)) = 0$.

We notice that the discontinuities of f are also discontinuities of g and g may have *more* discontinuities on ∂E, so we obtain

$$
D_g(I) = D_f(E) \cup D_g(\partial E).
$$

Since $E^\circ \cap (\partial E)^\circ = \varnothing$, Theorem 16.3(b) implies that

$$
\mu(D_g(I)) = \mu(D_f(E)) + \mu(D_g(\partial E)). \tag{16.20}
$$

Next, it is obvious that $D_g(\partial E) \subseteq \partial E$. Since E is Jordan measurable, Theorem 16.13 shows that $J(\partial E) = 0$. Applying Problems 16.2 and 16.3, we conclude that $\mu(D_g(\partial E)) = 0$. Hence it follows from the expression (16.20) that $\mu(D_g(I)) = 0$ if and only if $\mu(D_f(E)) = 0$. Hence our desired result follows directly from Definition 16.15 (Riemann Integrable on E). We have completed the proof of the problem. ∎

Problem 16.9

(⋆) *Prove Theorem 16.17(c).*

Proof. We remark that $\chi_E = \chi_{A \cup B} = \chi_A + \chi_B - \chi_{A \cap B}$, so we have

$$
\int_E f \, d\mathbf{x} = \int_{E \subseteq I} f \chi_E \, d\mathbf{x} = \int_I f \chi_A \, d\mathbf{x} + \int_I f \chi_B \, d\mathbf{x} - \int_I f \chi_{A \cap B} \, d\mathbf{x}. \tag{16.21}
$$

By Remark 16.2(b), since $A \cap B = \varnothing$, $f \chi_{A \cap B} = 0$ on I and then

$$
\int_I f \chi_{A \cap B} \, d\mathbf{x} = 0.
$$

By this, the expression (16.21) reduces to

$$
\int_E f \, d\mathbf{x} = \int_I f \chi_A \, d\mathbf{x} + \int_I f \chi_B \, d\mathbf{x}
$$

which is our required result, completing the proof of the problem. ∎

Problem 16.10

$(\star)$ Let $S = [0,1] \times [0,1]$. *Prove that*

$$\int_S y^3 e^{xy^2} \, \mathrm{d}(x,y) = \frac{e-2}{2}.$$

Proof. It is evident that f is Riemann integrable with respect to each variable. Furthermore, since $f(x,y) = y^3 e^{xy^2}$ is continuous on S, Theorem 16.10 (The Lebesgue's Integrability Condition) ensures that f satisfies the hypotheses of Fubini's Theorem. Hence we obtain

$$\int_S y^3 e^{xy^2} \, \mathrm{d}(x,y) = \int_0^1 \int_0^1 y^3 e^{xy^2} \, \mathrm{d}x \, \mathrm{d}y$$

$$= \int_0^1 y^3 \left(\int_0^1 e^{xy^2} \, \mathrm{d}x \right) \mathrm{d}y$$

$$= \int_0^1 y(e^{y^2} - 1) \, \mathrm{d}y$$

$$= \frac{e-2}{2}.$$

This completes the proof of the problem. $\blacksquare$

Problem 16.11

$(\star)$ *Suppose that* $f_k \in \mathcal{R}$ *on* $I_k = [a_k, b_k]$, *where* $k = 1, 2, \ldots, n$. *Verify that*

$$\int_I f_1(x_1) \cdots f_n(x_n) \, \mathrm{d}x_1 \cdots \mathrm{d}x_n = \left(\int_{a_1}^{b_1} f_1(x_1) \, \mathrm{d}x_1 \right) \times \cdots \times \left(\int_{a_n}^{b_n} f_n(x_n) \, \mathrm{d}x_n \right),$$

where $I = I_1 \times \cdots \times I_n$.

Proof. For each $k = 1, 2, \ldots, n$, let D_k be the set of discontinuities of f_k on I_k. Let D be the set of discontinuities of the function

$$f = f_1 f_2 \cdots f_n$$

on I. By Theorem 16.10 (The Lebesgue's Integrability Condition), we see that $\mu(D_k) = 0$, where $k = 1, 2, \ldots, n$. It is evident that

$$D \subseteq \bigcup_{k=1}^{n} I_1 \times \cdots \times I_{k-1} \times D_k \times I_{k+1} \times \cdots \times I_n.$$

Thus we know from Theorem 16.3(a) and then Definition 16.2 (The Measure of Intervals in $\mathbb{R}^n$) that

$$\mu(D) \leq \sum_{k=1}^{n} \mu(I_1 \times \cdots \times I_{k-1} \times D_k \times I_{k+1} \times \cdots \times I_n)$$

$$= \sum_{k=1}^{n} \mu(I_1) \times \cdots \times \mu(I_{k-1}) \times \mu(D_k) \times \mu(I_{k+1}) \times \cdots \times \mu(I_n)$$

$$= 0.$$

By Theorem 16.10 (The Lebesgue's Integrability Condition), it yields that $f \in \mathscr{R}$ on I. By repeated use of Fubini's Theorem, we establish that

$$
\begin{aligned}
\int_I f_1(x_1) \cdots f_n(x_n)\, dx_1 \cdots dx_n &= \int_{a_1}^{b_1} \left(\int_{I_2 \times \cdots \times I_k} f_1(x_1) \cdots f_n(x_n)\, dx_2 \cdots dx_n \right) dx_1 \\
&= \left(\int_{a_1}^{b_1} f_1(x_1)\, dx_1 \right) \times \left(\int_{I_2 \times \cdots \times I_k} f_2(x_2) \cdots f_n(x_n)\, dx_2 \cdots dx_n \right) \\
&= \left(\int_{a_1}^{b_1} f_1(x_1)\, dx_1 \right) \times \left(\int_{a_2}^{b_2} f_2(x_2)\, dx_2 \right) \\
&\quad \times \left(\int_{I_3 \times \cdots \times I_k} f_3(x_3) \cdots f_n(x_n)\, dx_3 \cdots dx_n \right) \\
&= \cdots \\
&= \left(\int_{a_1}^{b_1} f_1(x_1)\, dx_1 \right) \left(\int_{a_2}^{b_2} f_2(x_2)\, dx_2 \right) \cdots \left(\int_{a_n}^{b_n} f_n(x_n)\, dx_n \right).
\end{aligned}
$$

We have completed the proof of the problem.   ∎

Problem 16.12

(⋆) Suppose that $Q = [0,1] \times \cdots \times [0,1]$ and $\mathbf{y} = (1, 1, \ldots, 1)$. Prove that

$$\int_Q e^{-\mathbf{x} \cdot \mathbf{y}}\, d\mathbf{x} = \left(\frac{e-1}{e} \right)^n.$$

Proof. If $\mathbf{x} = (x_1, x_2, \ldots, x_n)$, then we have $-\mathbf{x} \cdot \mathbf{y} = -(x_1 + x_2 + \cdots + x_n)$ and thus

$$e^{-\mathbf{x} \cdot \mathbf{y}} = e^{-x_1} e^{-x_2} \cdots e^{-x_n}.$$

Since $e^{-x_k} \in \mathscr{R}$ on $[0,1]$, it deduces from Problem 16.11 that

$$\int_Q e^{-\mathbf{x} \cdot \mathbf{y}}\, d\mathbf{x} = \left(\int_0^1 e^{-x_1}\, dx_1 \right) \times \cdots \times \left(\int_0^1 e^{-x_n}\, dx_n \right) = \left(\frac{e-1}{e} \right)^n,$$

completing the proof of the problem.   ∎

Problem 16.13

(⋆) Let $a < A$ and $b < B$. Suppose that $f(x,y) = \dfrac{\partial^2}{\partial x \partial y} F(x,y)$ is continuous on $Q = [a, A] \times [b, B]$ and

$$I = \int_Q f(x,y)\, d(x,y).$$

Show that

$$I = F(A, B) - F(a, B) - F(A, b) + F(a, b).$$

Proof. Since f is continuous on Q, it satisfies all the hypotheses of Fubini's Theorem. Therefore, we have

$$I = \int_b^B \left(\int_a^A f(x,y) \, dx \right) dy. \tag{16.22}$$

By the Second Fundamental Theorem of Calculus, we know that

$$\int_a^A f(x,y) \, dx = \int_a^A \frac{\partial^2}{\partial x \partial y} F(x,y) \, dx = \frac{\partial}{\partial y} F(x,y) \Big|_a^A = \frac{\partial}{\partial y} F(A,y) - \frac{\partial}{\partial y} F(a,y). \tag{16.23}$$

Substituting the result (16.23) into the integral (16.22), we get

$$I = \int_b^B \left[\frac{\partial}{\partial y} F(A,y) - \frac{\partial}{\partial y} F(a,y) \right] dy = \int_b^B \frac{\partial}{\partial y} F(A,y) \, dy - \int_b^B \frac{\partial}{\partial y} F(a,y) \, dy.$$

Applying the Second Fundamental Theorem of Calculus again, we obtain immediately that

$$I = F(A,y) \Big|_b^B - F(a,y) \Big|_b^B = F(A,B) - F(A,b) - F(a,B) + F(a,b).$$

This completes the proof of the problem. ∎

Problem 16.14

(⋆) *Suppose that $S = [0,1] \times [0,1]$ and*

$$f(x,y) = \begin{cases} x + y - 1, & \text{if } x + y \le 1; \\ 0, & \text{otherwise.} \end{cases} \tag{16.24}$$

Prove that

$$\int_S f \, d(x,y) = -\frac{1}{6}.$$

Proof. Since f is continuous on S, it satisfies all the requirements of Fubini's Theorem. Thus we have

$$\int_S f \, d(x,y) = \int_0^1 \left(\int_0^1 f(x,y) \, dy \right) dx. \tag{16.25}$$

By the definition (16.24), the integral (16.25) reduces to

$$\begin{aligned}
\int_S f \, d(x,y) &= \int_0^1 \left(\int_0^{1-x} (x + y - 1) \, dy \right) dx \\
&= \int_0^1 \left(xy + \frac{y^2}{2} - y \right) \Big|_0^{1-x} \, dx \\
&= \int_0^1 \left[x(1-x) + \frac{(1-x)^2}{2} - (1-x) \right] dx \\
&= -\frac{1}{2} \int_0^1 (1-x)^2 \, dx \\
&= -\frac{1}{6},
\end{aligned}$$

completing the proof of the problem. ∎

Problem 16.15

$(\star)(\star)$ *Denote $S = [0,1] \times [0,1]$. Define $f : S \to \mathbb{R}$ by*

$$f(x,y) = \begin{cases} 0, & \text{if at least one of } x \text{ or } y \text{ is irrational;} \\ \frac{1}{n}, & \text{if } x, y \in \mathbb{Q} \text{ and } x = \frac{m}{n}, \end{cases} \tag{16.26}$$

where m and n are relatively prime and $n > 0$. Prove that

$$\int_0^1 f(x,y)\,\mathrm{d}x = \int_0^1 \left(\int_0^1 f(x,y)\,\mathrm{d}x \right) \mathrm{d}y = \int_S f(x,y)\,\mathrm{d}(x,y) = 0,$$

but $f(x,y) \notin \mathscr{R}$ on $[0,1]$ for every rational x.

Proof. On $([0,1] \setminus \mathbb{Q}) \times [0,1]$, we have $f(x,y)$ is continuous and zero. Given $\epsilon > 0$. Let $\{q_1, q_2, \ldots\} = \mathbb{Q} \cap [0,1]$ and $I_k = (q_k - \frac{\epsilon}{2^{k+1}}, q_k + \frac{\epsilon}{2^{k+1}})$. Then we have

$$\mathbb{Q} \cap [0,1] \subseteq \bigcup_{k=1}^{\infty} I_k \quad \text{and} \quad \sum_{k=1}^{\infty} \mu(I_k) = \sum_{k=1}^{\infty} \frac{\epsilon}{2^k} < \epsilon.$$

By the definition, $\mu(\mathbb{Q} \cap [0,1]) = 0$ and Theorem 16.10 (The Lebesgue's Integrability Condition) implies that $f(x,y) \in \mathscr{R}$ on S. If P is a partition of S into p rectangles $I_1, I_2, \ldots, I_p$. Since each rectangle must contain a point with irrational coordinates, the definition (16.26) shows that $L(P, f) = 0$ and then

$$\int_S f(x,y)\,\mathrm{d}(x,y) = 0 \tag{16.27}$$

by Theorem 16.8 (The Riemann Integrability Condition).

Next, if y is irrational, then $f(x,y) = 0$ for all $x \in [0,1]$. Therefore, Remark 16.2(b) shows that

$$\int_0^1 f(x,y)\,\mathrm{d}x = 0. \tag{16.28}$$

If $y \in \mathbb{Q} \cap [0,1]$, then we know from [18, Exercise 18, p. 100] that the function[d] $f(x,y)$ is continuous at every irrational x in $[0,1]$. Thus it follows from [25, Theorem 9.4, p. 159] that $f(x,y) \in \mathscr{R}$ on $[0,1]$ for every rational $y \in \mathbb{Q} \cap [0,1]$. By Theorem 14.6(a), we see that

$$\mathscr{R} \int_0^1 f(x,y)\,\mathrm{d}x = \int_{[0,1]} f(x,y)\,\mathrm{d}m = \int_{\mathbb{Q} \cap [0,1]} f(x,y)\,\mathrm{d}m + \int_{[0,1] \setminus \mathbb{Q}} f(x,y)\,\mathrm{d}m. \tag{16.29}$$

Since $m(\mathbb{Q} \cap [0,1]) = 0$ and $f(x,y) = 0$ on $[0,1] \setminus \mathbb{Q}$ by the definition (16.26), Theorem 14.3(f) ensures that the two Lebesgue integrals on the right-hand side of the equation (16.29) are zero. In other words, we have

$$\int_0^1 f(x,y)\,\mathrm{d}x = 0 \tag{16.30}$$

if $y \in \mathbb{Q} \cap [0,1]$. Now we combine the results (16.28) and (16.30) to conclude that

$$\int_0^1 f(x,y)\,\mathrm{d}x = 0 \tag{16.31}$$

---

[d]It is, in fact, the **Riemann function**, see [25, Problems 7.5 and 9.8, pp. 105, 106, 166]

for every $y \in [0,1]$ and so

$$\int_0^1 \left(\int_0^1 f(x,y) \, \mathrm{d}x \right) \mathrm{d}y = 0. \tag{16.32}$$

Hence our desired results follows immediately from the results (16.27), (16.31) and (16.32).

However, suppose that $x \in \mathbb{Q} \cap [0,1]$. Then we have $x = \frac{m}{n}$, where m and n are relatively prime and $n > 0$. Thus we get from the definition (16.26) that

$$f(x,y) = \begin{cases} 0, & \text{if } y \in [0,1] \setminus \mathbb{Q}; \\[2mm] \frac{1}{n}, & \text{if } y \in \mathbb{Q} \cap [0,1]. \end{cases} \tag{16.33}$$

Obviously, the function (16.33) is nowhere continuous because it is a multiple of the **Dirichlet function** $D(y)$.[e] Consequently, $f(x,y) \notin \mathscr{R}$ on $[0,1]$ for every fixed $x \in \mathbb{Q} \cap [0,1]$. This completes the proof of the problem. ∎

Remark 16.7

Problem 16.15 shows that the condition $f(\cdot, y) \in \mathscr{R}$ on $[c,d]$ for every $x \in [a,b]$ in Fubini's Theorem cannot be relaxed. In fact, one can find counterexamples to show that the other hypotheses cannot be omitted.

Problem 16.16

(⋆) *Suppose that* $E = \{(x,y,z) \in \mathbb{R}^3 \,|\, 0 \le x \le 1, \, 0 \le y \le 1 - x \text{ and } 0 \le z \le 1 - x - y\}$ *and* $f(x,y,z) = x$. *Prove that*

$$\int_E f \, \mathrm{d}\mathbf{x} = \frac{1}{24}.$$

Proof. We notice that if $F = \{(x,y) \in \mathbb{R}^2 \,|\, 0 \le x \le 1 \text{ and } 0 \le y \le 1 - x\}$, $\phi(x,y) = 0$ and $\varphi(x,y) = 1 - x - y$, then F is clearly a closed Jordan measurable subset of $\mathbb{R}^2$ (in fact, F is the area bounded by the straight lines $y = 1 - x$, $x = 0$ and $y = 0$) and both ϕ and φ are continuous on F. Thus Theorem 16.18 implies that

$$\int_E f \, \mathrm{d}\mathbf{x} = \iiint_E f(x,y,z) \, \mathrm{d}x \, \mathrm{d}y \, \mathrm{d}z = \int_F \left(\int_0^{1-x-y} x \, \mathrm{d}z \right) \mathrm{d}(x,y). \tag{16.34}$$

Similarly, if $I = [0,1]$, $\phi(x) = 0$ and $\varphi(x) = 1 - x$, then I is a closed Jordan measurable subset of $\mathbb{R}$ and both $\phi(x) = 0$ and $\varphi(x) = 1 - x$ are continuous on I. Therefore, we apply Theorem 16.18 again to the integral on the right-hand side of (16.34) to obtain

$$\begin{aligned} \int_E f \, \mathrm{d}\mathbf{x} &= \int_F \left(\int_0^{1-x-y} x \, \mathrm{d}z \right) \mathrm{d}(x,y) \\ &= \int_0^1 \int_0^{1-x} \int_0^{1-x-y} x \, \mathrm{d}z \, \mathrm{d}y \, \mathrm{d}x \\ &= \int_0^1 \int_0^{1-x} (x - x^2 - xy) \, \mathrm{d}y \, \mathrm{d}x \end{aligned}$$

---

[e]Read [25, Problems 7.2 and 7.11, pp. 103, 104, 110, 111]

$$= \frac{1}{2} \int_0^1 (x - 2x^2 + x^3) \, \mathrm{d}x$$

$$= \frac{1}{24},$$

completing the proof of the problem.　　　　　　　　　　　　　　　　　　　■

16.4　Applications of the Mean Value Theorem

Problem 16.17

$\bigstar$ $\bigstar$ *Suppose that* $E \subseteq \mathbb{R}^n$, $f \in \mathscr{R}$ *on* E *and*

$$\int_E f \, \mathrm{d}\mathbf{x} = 0. \tag{16.35}$$

Let $F = \{\mathbf{x} \in E \mid f(\mathbf{x}) < 0\}$ *and* $J(F) = 0$. *Prove that there corresponds a set* S *with* $\mu(S) = 0$ *and* $f(\mathbf{x}) = 0$ *for every* $\mathbf{x} \in E \setminus S$.

Proof. Since $J(F) = 0$, F is Jordan measurable and the particular case of the Mean Value Theorem for Multiple Integrals implies that

$$\int_F f \, \mathrm{d}\mathbf{x} = 0. \tag{16.36}$$

Besides, it follows from Theorem 16.14 that $E \setminus F$ is also Jordan measurable. By the conditions (16.35) and (16.36), and Theorem 16.17(c) (Properties of Integration on E), we obtain

$$\int_{E \setminus F} f \, \mathrm{d}\mathbf{x} = \int_E f \, \mathrm{d}\mathbf{x} - \int_F f \, \mathrm{d}\mathbf{x} = 0,$$

i.e., $f \in \mathscr{R}$ on $E \setminus F$. Next, if A is the set of the discontinuities of f in $E \setminus F$, then we deduce from Theorem 16.16 that $\mu(A) = 0$.

Take $\mathbf{p} \in E \setminus (F \cup A \cup \partial E)$. Since $\mathbf{p} \notin A$, f is continuous at $\mathbf{p}$. If $f(\mathbf{p}) \neq 0$, then $f(\mathbf{p}) > 0$. Since $\mathbf{p} \notin \partial E$, the continuity of f at $\mathbf{p}$ shows that there exists a $\delta > 0$ such that $N_\delta(\mathbf{p}) \subseteq E \setminus F$ and

$$f(\mathbf{x}) > \frac{f(\mathbf{p})}{2} \tag{16.37}$$

for all $\mathbf{x} \in N_\delta(\mathbf{p})$. By the definition, if $\mathbf{x} \in E \setminus F$, then we have $f(\mathbf{x}) \geq 0$. Note that $(E \setminus F) \setminus N_\delta(\mathbf{p})$ is Jordan measurable so that

$$\int_{(E \setminus F) \setminus N_\delta(\mathbf{p})} f \, \mathrm{d}\mathbf{x} \geq 0 \tag{16.38}$$

by Theorem 16.17(b) (Properties of Integration on E). In addition, we know from the inequalities (16.37) and (16.38) that

$$\int_{E \setminus F} f \, \mathrm{d}\mathbf{x} = \int_{N_\delta(\mathbf{p})} f \, \mathrm{d}\mathbf{x} + \int_{(E \setminus F) \setminus N_\delta(\mathbf{p})} f \, \mathrm{d}\mathbf{x} \geq \int_{N_\delta(\mathbf{p})} f \, \mathrm{d}\mathbf{x} > \mu(N_\delta(\mathbf{p})) \cdot \frac{f(\mathbf{p})}{2} > 0$$

which contradicts the hypothesis (16.35). In conclusion, we must have $f(\mathbf{p}) = 0$ and thus

$$f(\mathbf{x}) = 0 \qquad (16.39)$$

on $E \setminus (F \cup A \cup \partial E)$.

Finally, we claim that the set

$$S = F \cup A \cup \partial E$$

satisfies the requirements. By the result (16.39), it suffices to prove that $\mu(S) = 0$. Recall that $J(F) = 0$, so $\mu(F) = 0$ by Problem 16.2. By the properties of the boundary of a set,[f] we have

$$\partial E \subseteq \partial F \cup \partial(E \setminus F) \quad \text{and} \quad \partial(E \setminus F) = \overline{E \setminus F} \cap \overline{E \setminus (E \setminus F)} = \overline{E \setminus F} \cap \overline{F} = \overline{F}.$$

Consequently, we have

$$\partial E \subseteq \partial F \cup \overline{F}.$$

By Theorem 16.13, we have $J(\partial F) = 0$, so $\mu(\partial F) = 0$ by Problem 16.2. By Problem 16.1, we know that $J(\overline{F}) = J^*(\overline{F}) = J^*(F) = J(F) = 0$ which implies $\mu(\overline{F}) = 0$ by Problem 16.2 again. Using Remark 16.2(a) twice, we get

$$\mu(\partial E) = 0 \quad \text{and} \quad \mu(S) = 0$$

which proves the claim. Hence we complete the analysis of the problem. ∎

Problem 16.18

(⋆) Let E be a Jordan measurable set of $\mathbb{R}^n$, $f \in \mathscr{R}$ on E and $g : E \to \mathbb{R}$ be bounded. If $F \subseteq E$ satisfies $J(F) = 0$ and $g(\mathbf{x}) = f(\mathbf{x})$ on $E \setminus F$, prove that $g \in \mathscr{R}$ on E and

$$\int_E f \, d\mathbf{x} = \int_E g \, d\mathbf{x}.$$

Proof. Since $f = g$ on $E \setminus F$, it follows from Theorem 16.17(c) (Properties of Integration on E) and Remark 16.5(a) that

$$\int_E g \, d\mathbf{x} = \int_{E \setminus F} g \, d\mathbf{x} + \int_F g \, d\mathbf{x} = \int_{E \setminus F} g \, d\mathbf{x} + 0 = \int_{E \setminus F} f \, d\mathbf{x} + \int_F f \, d\mathbf{x} = \int_E f \, d\mathbf{x}.$$

This completes the proof of the problem. ∎

Problem 16.19

(⋆)(⋆) Suppose that E is an open connected Jordan measurable subset of $\mathbb{R}^n$, $f \in \mathscr{R}$ on E and f is continuous on E. Prove that there is a point $\mathbf{p} \in E$ such that

$$\int_E f \, d\mathbf{x} = f(\mathbf{p}) J(E). \qquad (16.40)$$

---

[f]Refer to [2, Exercises 3.51 and 3.52, p. 69].

Proof. Since $f \in \mathscr{R}$ on E, f is bounded on E. Therefore, both $M = \sup\{f(\mathbf{x}) \,|\, \mathbf{x} \in E\}$ and $m = \inf\{f(\mathbf{x}) \,|\, \mathbf{x} \in E\}$ are finite and $m \le f(\mathbf{x}) \le M$ on E. By the Mean Value Theorem for Multiple Integrals, we have

$$mJ(E) \le \int_E f \,\mathrm{d}\mathbf{x} \le MJ(E)$$

which means that there exists a $\lambda \in [m, M]$ such that

$$\int_E f \,\mathrm{d}\mathbf{x} = \lambda J(E). \tag{16.41}$$

If $\lambda = m$, then we must have $m = M$. Otherwise, one can find a point $\mathbf{a} \in E$ such that

$$m < f(\mathbf{a}) < M. \tag{16.42}$$

Since f is continuous on E, the Sign-preserving Property[g] implies that there exists a $\delta > 0$ such that $f(\mathbf{x}) > m$ on $N_\delta(\mathbf{a})$. However, Theorem 16.17 (Properties of Integration on E) and Remark 16.6 give

$$\begin{aligned}
\int_E f \,\mathrm{d}\mathbf{x} &= \int_{N_\delta(\mathbf{a})} f \,\mathrm{d}\mathbf{x} + \int_{E \setminus N_\delta(\mathbf{a})} f \,\mathrm{d}\mathbf{x} \\
&> \int_{N_\delta(\mathbf{a})} m \,\mathrm{d}\mathbf{x} + \int_{E \setminus N_\delta(\mathbf{a})} m \,\mathrm{d}\mathbf{x} \\
&= m \int_E \mathrm{d}\mathbf{x} \\
&= mJ(E) \tag{16.43}
\end{aligned}$$

which contradicts the formula (16.41). In other words, f is a constant function and in fact, $f(\mathbf{x}) = m$ on E. Hence the formula (16.40) holds trivially.

Next, if $\lambda = M$, then we also have $m = M$. Otherwise, the inequality (16.42) holds for another point $\mathbf{b} \in E$. Thus the previous analysis can be repeated to obtain the same contradiction (16.43). Now we may suppose that

$$m < \lambda < M.$$

Since f is continuous on E and E is connected, $f(E)$ is a connected[h] subset of $[m, M]$. In fact, we conclude from [25, Theorem 4.16, p. 31] that $f(E)$ is an interval. By the definitions of M and m, we certainly have

$$(m, M) \subseteq f(E) \subseteq [m, M].$$

These observations show that one can find a point $\mathbf{p} \in E$ such that

$$f(\mathbf{p}) = \lambda.$$

This completes the proof of the problem.   ∎

---

[g]In fact, we require its generalized version with $\mathbb{R}^n$ as the domain in [25, p. 112].

[h]See [25, Theorem 7.12, p. 100].

16.5 Applications of the Change of Variables Theorem

Problem 16.20

$(\star)$ *Evaluate*

$$\int_D \sin \sqrt{x^2 + y^2} \, d(x,y),$$

where $D = \{(x,y) \mid \pi^2 \le x^2 + y^2 \le 4\pi^2\}$.

Proof. Let $E = \{(r,\theta) \mid \pi \le r \le 2\pi \text{ and } 0 < \theta < 2\pi\} = [\pi, 2\pi] \times (0, 2\pi)$. Then E is obviously Jordan measurable because $J(\partial E) = 0$ and $\phi(E) = D$, where

$$\phi(r,\theta) = (r\cos\theta, r\sin\theta)$$

and D is an annulus. Next, we note that if $V = (\frac{\pi}{2}, \frac{5\pi}{2}) \times (-\frac{\pi}{2}, \frac{5\pi}{2})$, then V is an open set containing $\overline{E} = [\pi, 2\pi] \times [0, 2\pi]$ and does not intersect the θ-axis. See Figure 16.3 for detials.

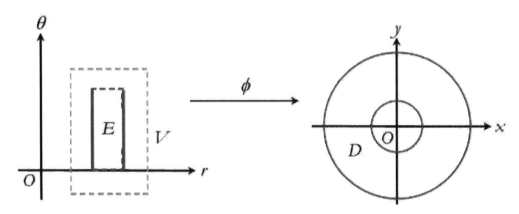

Figure 16.3: The mapping $\phi : E \to D$.

Finally, the function $f(x,y) = \sin\sqrt{x^2 + y^2}$ is continuous on $\mathbb{R}^2$ and $f(r\cos\theta, r\sin\theta) = \sin r$, so we must have $f \in \mathscr{R}$ on D and $f(r\cos\theta, r\sin\theta) \in \mathscr{R}$ on E. Hence we may apply the formula (16.9) to get

$$\int_D \sin\sqrt{x^2+y^2} \, dx \, dy = \int_E f(r\cos\theta, r\sin\theta) r \, dr \, d\theta$$

$$= \int_0^{2\pi} \int_\pi^{2\pi} r \sin r \, dr \, d\theta$$

$$= -\int_0^{2\pi} \left[r\cos r \Big|_\pi^{2\pi} - \int_\pi^{2\pi} \cos r \, dr \right] d\theta$$

$$= -\int_0^{2\pi} 3\pi \, d\theta$$

$$= -6\pi^2.$$

This completes the analysis of the problem. ∎

Problem 16.21

($\star$) *Evaluate*

$$\int_R (x+y)^3 \, \mathrm{d}x \, \mathrm{d}y,$$

where R is the parallelogram with vertices $(1,0), (3,1), (2,2)$ and $(0,1)$.

Proof. Consider $f(x,y) = (x+y)^3$ and $\phi : \mathbb{R}^2 \to \mathbb{R}^2$ defined by

$$\phi(u,v) = \left(\frac{2u+v}{3}, \frac{u-v}{3} \right).$$

Then it is continuously differentiable and injective on $\mathbb{R}^2$. In addition, if E is the parallelogram with vertices $(1,-2), (4,-2), (4,1)$ and $(1,1)$, then it is easily shown that $\phi(E) = D$, see Figure 16.4 below:

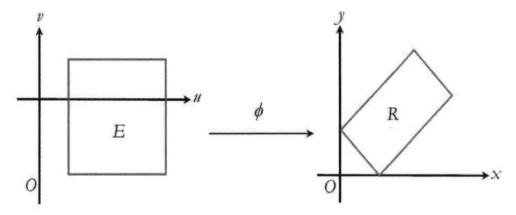

Figure 16.4: The mapping $\phi : E \to R$.

Since f is continuous on $\mathbb{R}^2$, we have $f \in \mathcal{R}$ on R by Theorem 16.16. Since $f(\phi(u,v)) = u^3$ which is continuous on $\mathbb{R}^2$, we have $f \circ \phi \in \mathcal{R}$ on E by Theorem 16.16. Finally, since E is clearly a Jordan measurable set, $\overline{E} \subseteq \mathbb{R}^2$ and

$$\det \mathbf{J}_\phi((u,v)) = \det \begin{pmatrix} \dfrac{2}{3} & \dfrac{1}{3} \\[2mm] \dfrac{1}{3} & -\dfrac{1}{3} \end{pmatrix} = -\frac{1}{3} \neq 0,$$

we obtain from the formula (16.8) that

$$\int_R (x+y)^3 \, \mathrm{d}x \, \mathrm{d}y = \int_{\phi(E)} (x+y)^3 \, \mathrm{d}x \, \mathrm{d}y = \int_E u^3 \cdot \frac{1}{3} \, \mathrm{d}u \, \mathrm{d}v = \frac{1}{3} \int_{-2}^{1} \int_{1}^{4} u^3 \, \mathrm{d}u \, \mathrm{d}v = \frac{255}{3},$$

completing the proof of the problem.   ■

Problem 16.22

($\star$) *Evaluate*

$$\int_D e^{-x^2-y^2}\, \mathrm{d}(x,y),$$

where $D = \{(x,y) \in \mathbb{R}^2 \,|\, x^2 + y^2 \le 1\}$.

Proof. Let $f(x,y) = e^{-x^2-y^2}$. Since it is continuous on D, $f \in \mathscr{R}$ on D by Theorem 16.16. Let $E = \{(r,\theta) \,|\, 0 \le r \le 1 \text{ and } 0 \le \theta < 2\pi\} = [0,1] \times [0, 2\pi)$. Then E is Jordan measurable because $J(\partial E) = 0$ and $\phi(E) = D$, where

$$\phi(r,\theta) = (r\cos\theta, r\sin\theta).$$

Using the polar coordinates (with the aid of Remark 16.4(b)), we have

$$
\begin{aligned}
\int_D e^{-x^2-y^2}\, \mathrm{d}(x,y) &= \int_{\phi(E)} e^{-x^2-y^2}\, \mathrm{d}(x,y) \\
&= \int_E re^{-r^2}\, \mathrm{d}r\, \mathrm{d}\theta \\
&= \int_0^{2\pi} \int_0^1 re^{-r^2}\, \mathrm{d}r\, \mathrm{d}\theta \\
&= -\frac{1}{2} \int_0^{2\pi} (e^{-1} - 1)\, \mathrm{d}\theta \\
&= \pi(1 - e^{-1}).
\end{aligned}
$$

This ends the proof of the problem.   ■

Problem 16.23

($\star$) *Suppose that E is a Jordan measurable subset of $\mathbb{R}^n$ and for every $\mathbf{x} \in E$, we have $-\mathbf{x} \in E$. Let f be an odd function such that $f \in \mathscr{R}$ on E. Prove that*

$$\int_E f\, \mathrm{d}\mathbf{x} = 0.$$

Proof. Define $\phi(\mathbf{x}) = -\mathbf{x}$. Then ϕ is continuously differentiable and injective on $\mathbb{R}^n$. In addition, we have $\phi(E) = E$ so that $f \in \mathscr{R}$ on $\phi(E)$. Therefore, we get $f \circ \phi \in \mathscr{R}$ on E. Finally, since $\overline{E} \subseteq \mathbb{R}^n$ and

$$\det \mathbf{J}_\phi(\mathbf{x}) = \det \begin{pmatrix} -1 & 0 & \cdots & 0 \\ 0 & -1 & \cdots & 0 \\ \vdots & \vdots & \ddots & 0 \\ 0 & 0 & \cdots & -1 \end{pmatrix} = (-1)^n \ne 0,$$

the integral in question satisfies all the requirements of the Change of Variables Theorem. Hence we deduce from the formula (16.8) and the fact $f(\phi(\mathbf{x})) = f(-\mathbf{x}) = -f(\mathbf{x})$ that

$$\int_E f(\mathbf{y}) \, d\mathbf{y} = \int_{\phi(E)} f(\mathbf{y}) \, d\mathbf{y} = \int_E f(\phi(\mathbf{x})) \cdot |\det \mathbf{J}_\phi(\mathbf{x})| \, d\mathbf{x} = -\int_E f(\mathbf{x}) \, d\mathbf{x}. \qquad (16.44)$$

Since the variable is dummy, the formula (16.45) implies that

$$\int_E f \, d\mathbf{x} = 0$$

which is our desired result. This completes the proof of the problem. ∎

Problem 16.24

$\bigstar$ $\bigstar$ *Show that*

$$\iint\limits_{0 < x < y < \pi} \ln |\sin(x - y)| \, dx \, dy = -\frac{1}{2}\pi^2 \ln 2.$$

Proof. Let $D = \{(x, y) \in \mathbb{R}^2 \,|\, 0 < x < y < \pi\}$ and $E = \{(u, v) \in \mathbb{R}^2 \,|\, 0 < u < v < 2\pi - u\}$. Define $\phi : \mathbb{R}^2 \to \mathbb{R}^2$ by

$$\phi(u, v) = \left(\frac{-u + v}{2}, \frac{u + v}{2} \right).$$

Then it is easy to see that $\phi(E) = D$, ϕ is continuously differentiable and injective on $\mathbb{R}^2$. Furthermore, E is Jordan measurable, $\overline{E} \subseteq \mathbb{R}^2$ and

$$\det \mathbf{J}_\phi((u, v)) = \det \begin{pmatrix} -\dfrac{1}{2} & \dfrac{1}{2} \\ \dfrac{1}{2} & \dfrac{1}{2} \end{pmatrix} = -\frac{1}{2} \neq 0.$$

Now the function $f(x, y) = \ln |\sin(x - y)|$ is continuous on D, so $f \in \mathscr{R}$ on D by Theorem 16.16. Since $f(\phi(u, v)) = \ln |\sin u|$ which is continuous on $\mathbb{R}^2$, Theorem 16.16 implies that $f \circ \phi \in \mathscr{R}$ on E. Thus it follows from the formula (16.8) that

$$\begin{aligned} I &= \iint\limits_D \ln |\sin(x - y)| \, dx \, dy \\ &= \frac{1}{2} \int_E \ln |\sin u| \, du \, dv \\ &= \frac{1}{2} \int_0^\pi \int_u^{2\pi - u} \ln |\sin u| \, dv \, du \\ &= \int_0^\pi (\pi - u) \ln |\sin u| \, du. \qquad (16.45) \end{aligned}$$

Next, we apply the substitution $y = \frac{\pi}{2} - u$ to the integral (16.45) to get

$$I = -\int_{\frac{\pi}{2}}^{-\frac{\pi}{2}} \left(\frac{\pi}{2} + y \right) \ln \left| \sin \left(\frac{\pi}{2} - y \right) \right| \, dy$$

$$= \int_{-\frac{\pi}{2}}^{\frac{\pi}{2}} \left(\frac{\pi}{2} + y\right) \ln|\cos y| \, dy$$

$$= \frac{\pi}{2} \int_{-\frac{\pi}{2}}^{\frac{\pi}{2}} \ln|\cos y| \, dy + \int_{-\frac{\pi}{2}}^{\frac{\pi}{2}} y \ln|\cos y| \, dy. \tag{16.46}$$

Since the first and the second integrands in the expression (16.46) are even and odd functions respectively, we are able to reduce it to

$$I = \pi \int_0^{\frac{\pi}{2}} \ln|\cos y| \, dy.$$

Notice that

$$\int_0^{\frac{\pi}{2}} \ln|\sin y| \, dy = \int_0^{\frac{\pi}{2}} \ln|\cos y| \, dy,$$

therefore we obtain

$$\frac{2}{\pi} I = 2 \int_0^{\frac{\pi}{2}} \ln|\cos y| \, dy$$

$$= \int_0^{\frac{\pi}{2}} \ln|\cos y| \, dy + \int_0^{\frac{\pi}{2}} \ln|\cos y| \, dy$$

$$= \int_0^{\frac{\pi}{2}} \ln|\cos y| \, dy + \int_0^{\frac{\pi}{2}} \ln|\sin y| \, dy$$

$$= \int_0^{\frac{\pi}{2}} \ln\left|\frac{2 \sin y \cos y}{2}\right| \, dy$$

$$= \frac{1}{2} \int_0^{\frac{\pi}{2}} \ln|\sin 2y| \, d(2y) - \int_0^{\frac{\pi}{2}} \ln 2 \, dy$$

$$= \frac{I}{\pi} - \frac{\pi}{2} \ln 2$$

which implies that

$$I = -\frac{\pi^2}{2} \ln 2.$$

We have completed the proof of the problem. ∎

Problem 16.25

(⋆) Prove that

$$\int_V (x^2 + y^2 + z^2) \, d(x, y, z) = \frac{4\pi}{5},$$

where $V = \{(x, y, z) \mid x^2 + y^2 + z^2 = 1\}$.

Proof. If $E = \{(r, \varphi, \theta) \mid 0 \le r \le 1,\ 0 \le \varphi \le \pi \text{ and } 0 \le \theta \le 2\pi\}$, then we have $\phi(E) = V$, where

$$\phi(r, \varphi, \theta) = (r \sin\varphi \cos\theta, r \sin\varphi \sin\theta, r \cos\theta).$$

Since $\det \mathbf{J}_\phi((r, \varphi, \theta)) = r^2 \sin \varphi$, we know from the spherical coordinates (with the aid of Remark 16.4(b)) that

$$
\begin{aligned}
\int_V (x^2 + y^2 + z^2) \, \mathrm{d}(x, y, z) &= \int_{\phi(E)} (x^2 + y^2 + z^2) \, \mathrm{d}(x, y, z) \\
&= \int_E r^2 \cdot r^2 \sin \varphi \, \mathrm{d}r \, \mathrm{d}\varphi \, \mathrm{d}\theta \\
&= \int_0^1 \int_0^\pi \int_0^{2\pi} r^4 \sin \varphi \, \mathrm{d}r \, \mathrm{d}\varphi \, \mathrm{d}\theta.
\end{aligned}
\tag{16.47}
$$

Applying Problem 16.11 to the integral (16.47), we establish immediately that

$$
\int_V (x^2 + y^2 + z^2) \, \mathrm{d}(x, y, z) = \int_0^1 r^4 \, \mathrm{d}r \int_0^\pi \sin \varphi \, \mathrm{d}\varphi \int_0^{2\pi} \mathrm{d}\theta = \frac{4\pi}{5},
$$

completing the proof of the problem.   ∎

Problem 16.26

(★)(★) Given that f is a differentiable function and

$$
F(t) = \int_{D(t)} f(x^2 + y^2 + z^2) \, \mathrm{d}(x, y, z),
$$

where $D(t) = \{(x, y, z) \mid x^2 + y^2 + z^2 \le t^2\}$. Prove that

$$
F'(t) = 4\pi t^2 f(t^2).
$$

Proof. Let $E(t) = \{(r, \varphi, \theta) \mid 0 \le r \le t, \, 0 \le \varphi \le \pi \text{ and } 0 \le \theta \le 2\pi\}$. Then it is easy to see that

$$
\phi(E(t)) = D(t),
$$

where

$$
\phi(r, \varphi, \theta) = (r \sin \varphi \cos \theta, r \sin \varphi \sin \theta, r \cos \theta).
$$

Hence we follow from the formula (16.8) (with the aid of Remark 16.4(b)) and Problem 16.11 that

$$
\begin{aligned}
F(t) &= \int_{\phi(E(t))} f(r^2) r^2 \, \mathrm{d}r \, \mathrm{d}\varphi \, \mathrm{d}\theta \\
&= \int_0^t \int_0^\pi \int_0^{2\pi} r^2 f(r^2) \, \mathrm{d}r \, \mathrm{d}\varphi \, \mathrm{d}\theta \\
&= \int_0^{2\pi} \mathrm{d}\theta \int_0^\pi \sin \varphi \, \mathrm{d}\varphi \int_0^t r^2 f(r^2) \, \mathrm{d}r \\
&= 4\pi \int_0^t r^2 f(r^2) \, \mathrm{d}r.
\end{aligned}
$$

Since f is differentiable, we deduce from the First Fundamental Theorem of Calculus that

$$
F'(t) = 4\pi t^2 f(t^2),
$$

completing the proof of the problem.   ∎

Index

Bibliography

[1] T. M. Apostol, *Calculus Vol. 1*, 2nd ed., John Wiley & Sons, Inc., 1967.

[2] T. M. Apostol, *Mathematical Analysis*, 2nd ed., Addison-Wesley Publishing Company, 1974.

[3] A. G. Aksoy and M. A. Khamsi, *A Problem Book in Real Analysis*, Springer-Verlag, New York, 2009.

[4] R. G. Bartle and D. R. Sherbert, *Introduction to Real Analysis*, 4th ed., John Wiley & Sons, Inc., 2011.

[5] G. B. Folland, *Real Analysis: Modern Techniques and Their Applications*, 2nd ed., New York: Wiley, 1999.

[6] M. Hata, *Problems and Solutions in Real Analysis*, Hackensack, NJ: World Scientific, 2007.

[7] E. W. Hobson, On the Second Mean-Value Theorem of the Integral Calculus, *Proc. Lond. Math. Soc.* Ser. 2, Vol. 7, pp. 14 - 23, 1909.

[8] S. G. Krantz and H. R. Parks, *The Implicit Function Theorem: History, Theory, and Applications*, Boston: Birkhäuser, 2002.

[9] F. Jones, *Lebesgue Integration on Euclidean Space*, Rev. ed., Boston: Jones and Bartlett, 2001.

[10] David C. Lay, *Linear Algebra and Its Applications*, 4th ed., Addison-Wesley Publishing Company, 2012.

[11] P. A. Loeb and E. Talvila, Lusin's Theorem and Bochner Integration, *Scientiae Mathematicae Japonicae*, Vol. 10, pp. 55 - 62, 2004.

[12] J. R. Munkres, *Topology*, 2nd ed., Upper Saddle River, N.J.: Prentice-Hall, 2000.

[13] P. J. Olver and C. Shakiban, Applied Linear Algebra, 2nd ed., Cham: Springer International Publishing, 2018.

[14] O. de Oliveira, The Implicit and Inverse Function Theorems: Easy Proofs, *Real Anal. Exchange*, Vol. 39, No. 1, pp. 207 - 218, 2013.

[15] T.-L. T. Rădulescu, V. D. Rădulescu and T. Andreescu, *Problems in Real Analysis: Advanced Calculus on the Real Axis*, Springer-Verlag, New York, 2009.

[16] I. M. Roussos, *Improper Riemann Integrals*, Boca Raton: CRC Press, 2014.

[17] H. L. Royden and P. M. Fitzpatrick, *Real Analysis*, 4th ed., Boston: Prentice Hall, 2010.

[18] W. Rudin, *Principles of Mathematical Analysis*, 3rd ed., Mc-Graw Hill Inc., 1976.

[19] W. Rudin, *Real and Complex Analysis*, 3rd ed., Mc-Graw Hill Inc., 1987.

[20] R. Shakarchi, *Problems and Solutions for Undergraduate Analysis*, Springer-Verlag, New York, 1998.

[21] E. M. Stein and R. Shakarchi, *Real Analysis: Measure Theorey, Integration and Hilbert Spaces*, Princeton, N. J.: Princeton University Press, 2005.

[22] T. Tao, *Analysis II*, 3rd ed., Singapore: Springer, 2016.

[23] T. Tao, *An Introduction to Measure Theory*, Providence, R.I.: American Mathematical Society, 2011.

[24] W. R. Wade, *An Introduction to Analysis*, 3rd, ed., Upper Saddle River: Pearson Prentice Hall, 2004.

[25] K. W. Yu, *Problems and Solutions for Undergraduate Real Analysis I*, Amazon.com, 2018.

[26] W. P. Ziemer, *Modern Real Analysis*, 2nd ed., Cham: Springer International Publishing, 2017.

[27] V. A. Zorich, *Mathematical Analysis I*, 2nd ed., Berlin: Springer, 2016.

[28] V. A. Zorich, *Mathematical Analysis II*, 2nd ed., Berlin: Springer, 2016.